Everyday Mathematics®

Student Math Journal 1

The University of Chicago
School Mathematics Project

The McGraw·Hill Companies

UCSMP Elementary Materials Component
Max Bell, Director

Authors
Max Bell
John Bretzlauf
Amy Dillard
Robert Hartfield
Andy Isaacs
James McBride, Director
Kathleen Pitvorec
Peter Saecker
Robert Balfanz*
William Carroll*

Technical Art
Diana Barrie

**First Edition only*

Photo Credits
Wally McNamee/Corbis, p. 140

Contributors
Tammy Belgrade, Diana Carry, Debra Dawson, Kevin Dorken, James Flanders, Laurel Hallman, Ann Hemwall, Elizabeth Homewood, Linda Klaric, Lee Kornhauser, Judy Korshak-Samuels, Deborah Arron Leslie, Joseph C. Liptak, Sharon McHugh, Janet M. Meyers, Susan Mieli, Donna Nowatzki, Mary O'Boyle, Julie Olson, William D. Pattison, Denise Porter, Loretta Rice, Diana Rivas, Michelle Schiminsky, Sheila Sconiers, Kevin J. Smith, Theresa Sparlin, Laura Sunseri, Kim Van Haitsma, John Wilson, Mary Wilson, Carl Zmola, Theresa Zmola

This material is based upon work supported by the National Science Foundation under Grant No. ESI-9252984. Any opinions, findings, and conclusions or recommendations expressed in this material are those of the authors and do not necessarily reflect the views of the National Science Foundation.

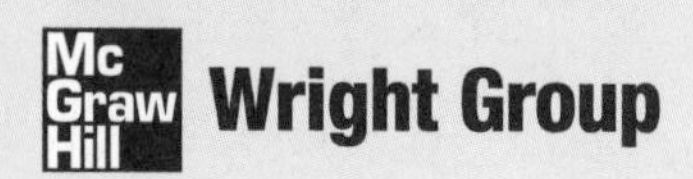

Send all inquiries to:
Wright Group/McGraw-Hill
P.O. Box 812960
Chicago, IL 60681

Printed in the United States of America.

ISBN 0-07-600035-4

4 5 6 7 8 9 10 11 12 DBH 10 08 07 06 05 04

The McGraw·Hill Companies

Contents

Unit 1: Number Theory

A note at the bottom of each journal page indicates when that page is first used.
Some pages will be used again during the course of the year.

Unit 2: Estimation and Computation

Unit 3: Geometry Explorations and the American Tour

Unit 4: Division

Unit 5: Fractions, Decimals, and Percents

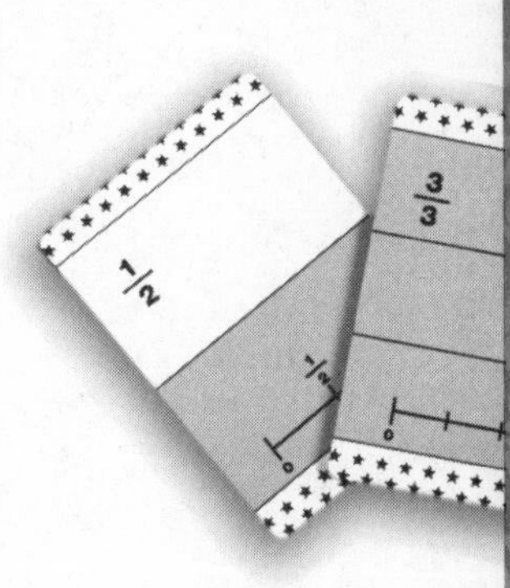

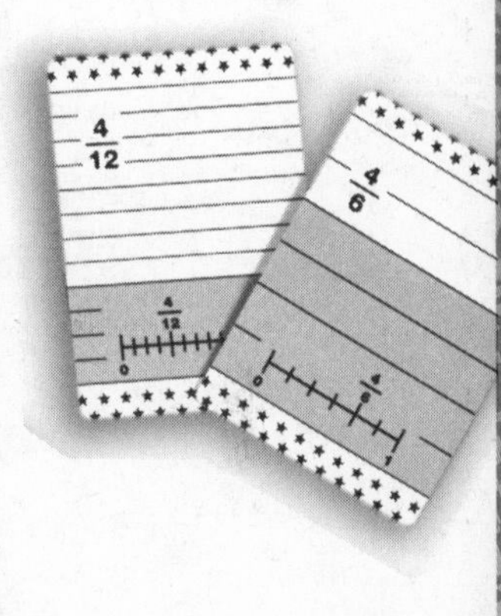

Unit 6: Using Data: Addition and Subtraction of Fractions

Welcome to *Fifth Grade Everyday Mathematics*

Much of what you learned in the first few years of *Everyday Mathematics* was basic training in mathematics and its uses. This year, you will practice and extend the skills and ideas you have learned. But you will also study more new ideas in mathematics—some of which your parents and older siblings may not have learned until high school! The authors of *Everyday Mathematics* believe that fifth graders in the 2000s can learn more and do more than people thought was possible 10 or 20 years ago.

Here are some of the things you will be asked to do in *Fifth Grade Everyday Mathematics:*

- Practice and extend your number sense, measure sense, and estimation skills.
- Review and extend your arithmetic, calculator, and thinking skills. You will work with fractions, decimals, percents, large whole numbers, and negative numbers.
- Continue your work with algebra, using variables in place of numbers.
- Refine your understanding of geometry. You will define and classify geometric figures more carefully than before. You will construct and transform figures. You will find the areas of 2-dimensional figures and volumes of 3-dimensional figures.
- Embark on the American Tour. You will study data about the history, people, and environment of the United States. You will learn to use and interpret many kinds of maps, graphs, and tables.
- Do many probability and statistics explorations with numerical data. You will use data that comes from questionnaires and experiments.

This year's activities will help you appreciate the beauty and usefulness of mathematics. We hope you will enjoy *Fifth Grade Everyday Mathematics.* We want you to become better at using mathematics, so that you may better understand the world you live in.

Student Reference Book Scavenger Hunt

Solve the problems on this page and on the next two pages. Use your *Student Reference Book* to help you.

Also, record where to find information in the *Student Reference Book* for each problem. You may not need to look for help in the *Student Reference Book,* but you will earn additional points for telling where you would look if you needed to.

When the class goes over the answers, keep score as follows:

- Give yourself **3 points** for each correct answer to a problem.
- Give yourself **5 points** for each correct page number in the *Student Reference Book.*

	Problem Points	Page Points
1. Circle the prime numbers in the following list: 1 2 6 9 13 20 31 63 72 *Student Reference Book,* page ________	________	________
2. Circle the composite numbers in the following list: 1 2 6 9 13 20 31 63 72 *Student Reference Book,* page ________	________	________
3. 5 meters = ________ centimeters *Student Reference Book,* page ________	________	________
4. 300 mm = ________ cm *Student Reference Book,* page ________	________	________
5. What is the perimeter of this figure? ________ ft *Student Reference Book,* page ________	________	________

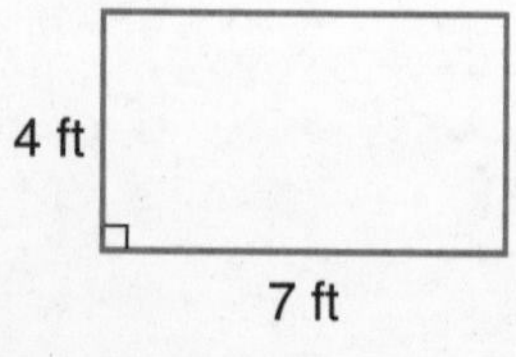

Date Time

Student Reference Book Scavenger Hunt (cont.)

	Problem Points	Page Points

6. 3 tablespoons = ________ teaspoons ________ ________

Student Reference Book, page ________

7. Is angle *RST* acute or obtuse? ________________ ________ ________

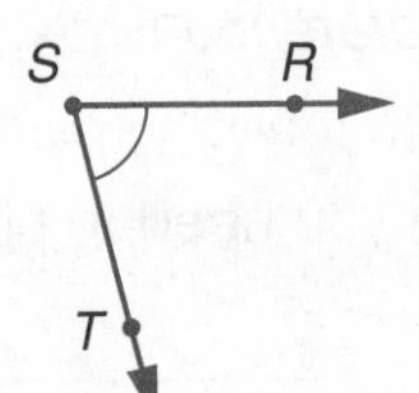

How can you tell? ________________

____

Student Reference Book, page ________

8. Rosie had the following scores on her spelling tests last month: 95, 87, 100, 92, and 78. ________ ________

What is the mean (or average) of these scores? ________

Student Reference Book, page ________

9. 23 ∗ 37 = ________ ________ ________

Student Reference Book, page ________

10. 369 + 1,347 = ____________ ________ ________

Student Reference Book, page ________

11. a. Is 73,491 divisible by 3? ____________ ________ ________

b. How can you tell without actually dividing? ________

____

____

____

____

Student Reference Book, page ________

Date Time

Student Reference Book Scavenger Hunt (cont.)

	Problem Points	Page Points
12. Name two fractions equivalent to $\frac{4}{6}$.	_______	_______
13. What materials do you need to play *Estimation Squeeze?*	_______	_______
14. What is the definition of a scalene triangle?	_______	_______
15. Use your calculator to find the square root of 9. _______	_______	_______

12. Name two fractions equivalent to $\frac{4}{6}$.

_______ and _______

Student Reference Book, page _______

13. What materials do you need to play *Estimation Squeeze?*

____

____

Student Reference Book, page _______

14. What is the definition of a scalene triangle?

____

____

____

Student Reference Book, page _______

15. Use your calculator to find the square root of 9. _______

Record the key sequence you used.

____

____

Student Reference Book, page _______

Total Problem Points _______

Total Page Points _______

Total Points _______

Date Time

Math Boxes 1.1

1. a. Write a 7-digit numeral that has
7 in the ones place,
8 in the millions place,
4 in the ten-thousands place,
and 0 in all other places.

___, ___ ___ ___, ___ ___ ___

b. Write this numeral in words.

____

____

____

2. Write each of the following in dollars-and-cents notation.

a. 5 dimes = ____________

b. 7 quarters = ____________

c. 10 quarters = ____________

d. 12 nickels = ____________

e. 18 dimes = ____________

3. Solve.

a. $982 + 497$

b. $384 + 499$

c. $125 + 47$

d. $958 + 1{,}003$

e. $271 + 634$

f. $367 + 548$

SRB 13 14

4. Below are a trapezoid, a rhombus, and a rectangle. Label each one.

____

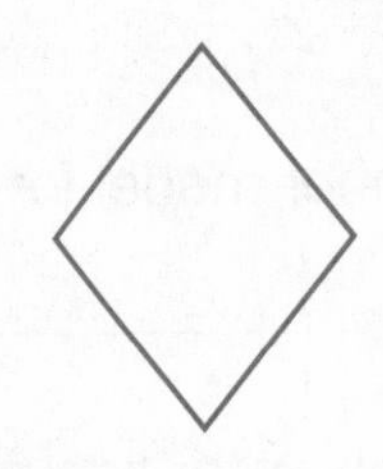

____

____

Date Time

Arrays

A **rectangular array** is an arrangement of objects into rows and columns. Each row has the same number of objects and each column has the same number of objects.

We can write a multiplication **number model** to describe a rectangular array.

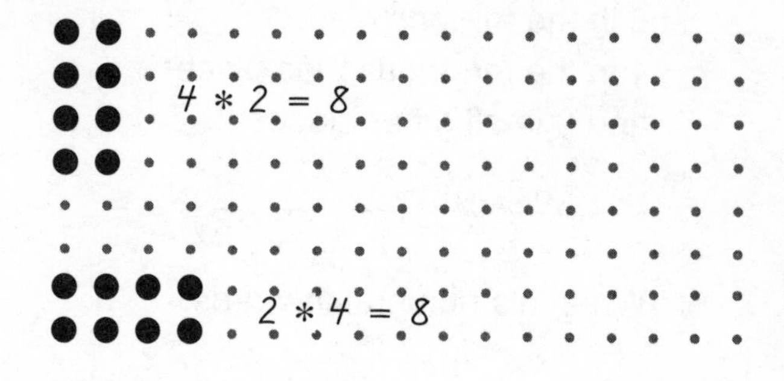

This is an array of 8 dots.
It has 4 rows with 2 dots in each row.
It has 2 columns with 4 dots in each column.

This is another array of 8 dots.
It has 2 rows with 4 dots in each row.
It has 4 columns with 2 dots in each column.

1. a. Take 10 counters. Make as many different rectangular arrays as you can using all 10 counters.

   b. Draw each array on the grid at the right by marking dots.

   c. Write the number model next to each array.

2. a. How many dots are in the array at the right?

   ____

   b. Write a number model for the array.

   ____

   c. Make as many other arrays as you can with the same number of counters as used in the array above. Draw each array on the grid at the right. Write a number model for each array.

Date Time

Math Boxes 1.2

1. Marcus drew 8 cards from a pile:
10, 8, 4, 5, 8, 6, 12, and 1.
Find the following landmarks.

a. Maximum ________

b. Minimum ________

c. Range ________

d. Median ________

2. Name five numbers between 0 and 1.

____

____

____

____

____

3. a. Make an array for the number sentence
$5 * 6 = 30$.

b. Write a number story for the number sentence.

____

____

____

4. a. Write the largest number you can make using each of the digits 7, 1, 0, 2, and 9 just once.

____

b. Write the smallest number. (It may not start with 0.)

____

5. Draw a line from each spinner to the number that best describes it.

$\frac{1}{3}$ $\frac{1}{4}$ 0.75 50%

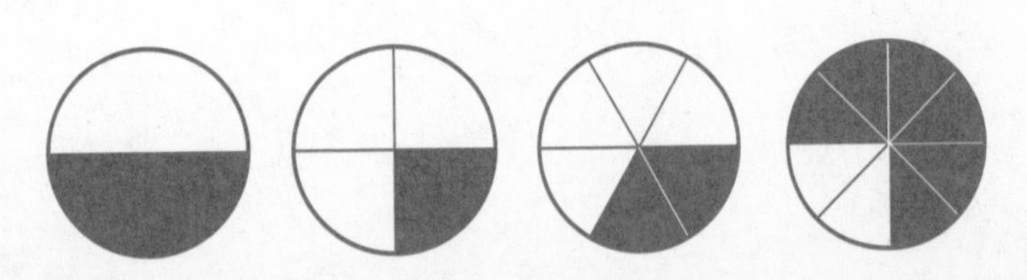

Date Time

Multiplication Facts Master List

Make a check mark next to each fact you missed and need to study.
Once you have learned a fact, write "OK" next to the check mark.

3s	5s	7s	9s
3 * 3	5 * 3	7 * 3	9 * 3
3 * 4	5 * 4	7 * 4	9 * 4
3 * 5	5 * 5	7 * 5	9 * 5
3 * 6	5 * 6	7 * 6	9 * 6
3 * 7	5 * 7	7 * 7	9 * 7
3 * 8	5 * 8	7 * 8	9 * 8
3 * 9	5 * 9	7 * 9	9 * 9
		7 * 10	9 * 10

4s	6s	8s	10s
4 * 3	6 * 3	8 * 3	10 * 3
4 * 4	6 * 4	8 * 4	10 * 4
4 * 5	6 * 5	8 * 5	10 * 5
4 * 6	6 * 6	8 * 6	10 * 6
4 * 7	6 * 7	8 * 7	10 * 7
4 * 8	6 * 8	8 * 8	10 * 8
4 * 9	6 * 9	8 * 9	10 * 9
	6 * 10	8 * 10	10 * 10

Date Time

Factor Pairs

A 2-row-by-5-column array

$2 * 5 = 10$

Factors Product

$2 * 5 = 10$ is a number model for the 2-by-5 array.
10 is the **product** of 2 and 5.
2 and 5 are whole-number **factors** of 10.
2 and 5 are a **factor pair** for 10.
1 and 10 are also factors of 10 because $1 * 10 = 10$.
1 and 10 are another **factor pair** for 10.

1. a. Use counters to make all possible arrays for the number 14.

   b. Write a number model for each array you make.

   ____

   ____

   c. List all the whole-number factors of 14.

   ____

2. Write number models to help you find all the factors of each number below.

Number	Number Models with 2 Factors	All Possible Factors
20		
16		
13		
27		
32		

Date Time

Math Boxes 1.3

1. a. Write a 6-digit numeral with
4 in the hundreds place,
8 in the hundred-thousands place,
3 in the ones place,
and 7s in all other places.

___ ___ ___, ___ ___ ___

b. Write this numeral in words.

____

____

____

____

____

2. Write each of the following in dollars-and-cents notation.

a. 12 dimes = ________

b. 12 quarters = ________

c. 15 nickels = ________

d. 3 quarters and 4 dimes = ________

e. 7 quarters and 3 nickels = ________

3. Add. Show your work.

a. 127 + 250 + 63 = ________

b. 67 + 109 + 318 = ________

c. 56 + 89 + 18 = ________

d. 39 + 71 + 177 = ________

4. a. Circle all of the quadrangles below.

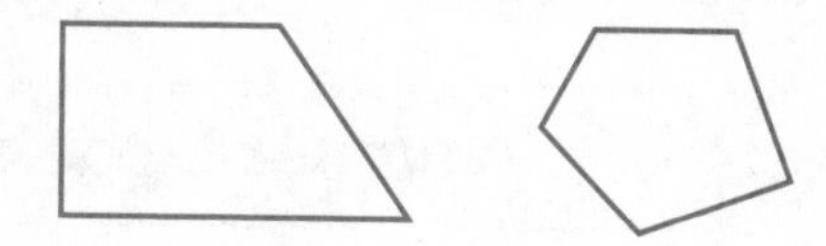

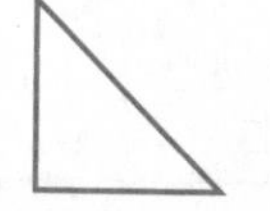

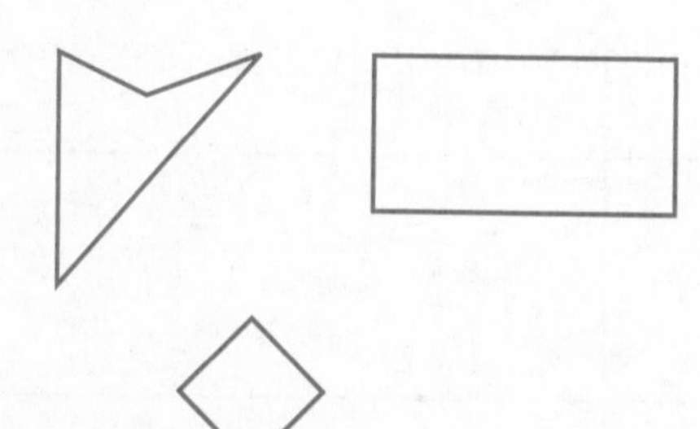

b. Put an X through each quadrangle that has one or more right angles.

Date Time

Math Boxes 1.4

1. Find the following landmarks for the set of numbers 28, 17, 45, 32, 29, 28, 14, 27.

a. Maximum ________

b. Minimum ________

c. Range ________

d. Median ________

2. Write five positive numbers that are less than 2.5.

____

____

____

____

____

3. a. Make an array for the number sentence $4 * 8 = 32$.

b. Write a number story for the number sentence.

____

____

____

4. a. What is the smallest whole number you can make using each of the digits 5, 8, 2, 7, and 4 just once?

____

b. What is the largest?

____

5. Draw a line from each spinner to the number that best describes it.

$66\frac{2}{3}\%$ $\frac{1}{2}$ 0.625 $\frac{2}{8}$

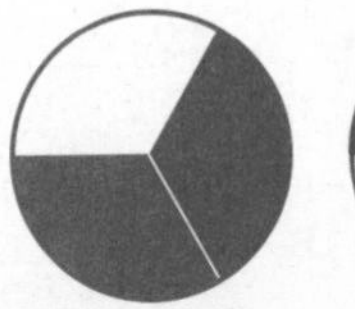

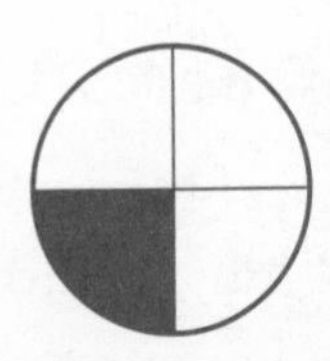
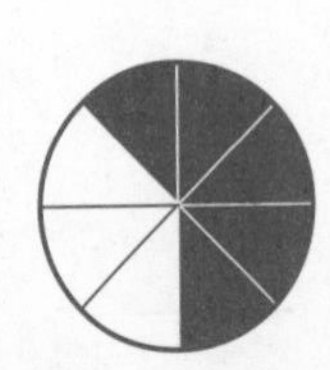

Date Time

Divisibility

Math Message

1. Circle the numbers that are divisible by 2.

28 57 33 112 123,456 211 5,374 900 399 705

2. True or false?

a. Even numbers end in 0, 2, 4, 6, or 8. ________

b. Even numbers are divisible by 2. ________

c. Every even number has 2 as a factor. ________

Suppose you divide a whole number by a second whole number. The answer may be a whole number or it may be a number that has a decimal part. If the answer is a whole number, we say that the first number is **divisible** by the second number. If it has a decimal part, the first number is *not* divisible by the second number.

Symbols		
3 ∗ 4	12 / 3	$\frac{12}{3}$
3 × 4	12 ÷ 3	$3\overline{)12}$

Example Is 135 divisible by 5?
To find out, divide 135 by 5.

135 / 5 = 27

The answer, 27, is a whole number. So 135 is divisible by 5.

Example Is 122 divisible by 5?
To find out, divide 122 by 5.

122 / 5 = 24.4

The answer, 24.4, has a decimal part. So 122 is *not* divisible by 5.

Use your calculator to help you answer these questions.

3. Is 267 divisible by 9? ________

4. Is 552 divisible by 6? ________

5. Is 809 divisible by 7? ________

6. Is 7,002 divisible by 3? ________

7. Is 4,735 divisible by 5? ________

8. Is 21,733 divisible by 4? ________

9. Is 5,268 divisible by 22? ________

10. Is 2,072 divisible by 37? ________

Date Time

Divisibility Tests

For many numbers, even large ones, it is possible to test for divisibility without actually dividing.

Here are the most useful divisibility tests:

- All numbers are **divisible by 1.**
- All even numbers (ending in 0, 2, 4, 6, or 8) are **divisible by 2.**
- A number is **divisible by 3** if the sum of its digits is divisible by 3.
  Example 246 is divisible by 3 because 2 + 4 + 6 = 12, and 12 is divisible by 3.
- A number is **divisible by 6** if it is divisible by both 2 and 3.
  Example 246 is divisible by 6 because it is divisible by 2 and by 3.
- A number is **divisible by 9** if the sum of its digits is divisible by 9.
  Example 51,372 is divisible by 9 because 5 + 1 + 3 + 7 + 2 = 18, and 18 is divisible by 9.
- A number is **divisible by 5** if it ends in 0 or 5.
- A number is **divisible by 10** if it ends in 0.

1. Test each number below for divisibility. Then check on your calculator.

	Divisible...					
Number	**by 2?**	**by 3?**	**by 6?**	**by 9?**	**by 5?**	**by 10?**
75		✓			✓	
7,960						
384						
3,725						
90						
36,297						

2. Find a 3-digit number that is divisible by both 3 and 5.

____

3. Find a 4-digit number that is divisible by both 6 and 9.

____

Date Time

Math Boxes 1.5

1. Complete.

a. 70 * 800 = ____________

b. 400 * 5,000 = ____________

c. 6,300 = ____________ * 90

d. 21,000 = 70 * ____________

e. 720,000 = 800 * ____________

2. a. Pencils are packed 18 to a box. How many pencils are there in 9 boxes?

____________ (unit)

b. Explain how you solved the problem.

____

____

3. Complete the table.

Fraction	Decimal	Percent
$\frac{3}{5}$		
		25%
	0.50	
$\frac{7}{10}$		
$\frac{85}{100}$		85%

4. a. Write a 5-digit numeral with
5 in the hundredths place,
8 in the tens place,
0 in the ones place,
3 in the thousandths place,
and 4 in the tenths place.

___ ___ . ___ ___ ___

b. Write this numeral in words.

____

____

SRB 30 31

5. Circle the numbers below that are divisible by 3.

221 381 474 922 726

6. Round 3,045,832 to the nearest …

a. million. ____________

b. thousand. ____________

c. ten-thousand. ____________

SRB 4 227

Date Time

Prime and Composite Numbers

A **prime number** has exactly two factors—1 and the number itself.
A **composite number** has more than two factors.

1. List all the factors of each number in the table. Write P if it is a prime number or C if it is a composite number.

Number	Factors	P or C	Number	Factors	P or C
2			21		
3	1, 3	P	22		
4			23		
5			24		
6	1, 2, 3, 6	C	25	1, 5, 25	C
7			26		
8			27		
9			28		
10			29		
11			30		
12			31		
13			32		
14			33		
15			34		
16			35		
17			36		
18			37		
19	1, 19	P	38		
20			39		

2. How many factors does each prime number have? ______________

3. Can a composite number have exactly 2 factors? ______________

   If yes, give an example of such a composite number. ______________

Date Time

Factor Captor Strategies

Work alone to answer the questions below. Then compare your answers with your partner's. If your answers don't agree with your partner's answers, try to convince your partner that your answers are correct.

1	2	3	4	5	6	7	8	9	10
11	12	13	14	15	16	17	18	19	20
21	22	23	24	25	26	27	28	29	30

1. Suppose you played *Factor Captor* using the above number grid. No numbers have been covered yet. Which is the best number choice you could make? Why?

____

____

____

2. Suppose that the 29 and 1 squares have already been covered. Which is the best number choice you could make? Why?

____

____

____

3. Suppose that the 29, 23, and 1 squares have already been covered. Which is the best number choice you could make? Why?

____

____

____

Date Time

Math Boxes 1.6

1. Write $<$ or $>$.

   a. 0.5 ________ 1.0

   b. 3.2 ________ 3.02

   c. 4.83 ________ 4.8

   d. 6.25 ________ 6.4

   e. 0.7 ________ 0.07

SRB 9 32 33

2. Round each number to the nearest ten-thousand.

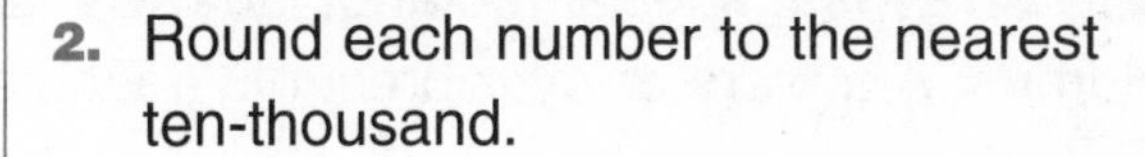

   a. 92,856 ____________

   b. 108,325 ____________

   c. 5,087,739 ____________

   d. 986,402 ____________

   e. 397,506 ____________

3. Subtract. Show your work.

   a. $105 - 59 =$ ____________

   b. $2,005 - 189 =$ ____________

   c. $680 - 74 =$ ____________

   d. $3,138 - 809 =$ ____________

4. List all of the factors of 36.

   ____

   ____

SRB 10 12

5. Math class is dismissed at 2:20 P.M. It is 1:53 P.M. How many more minutes before math class is dismissed?

   ____
   (unit)

Square Numbers

A **square array** is a special rectangular array that has the same number of rows as it has columns. A square array represents a whole number, called a **square number.**

The first four square numbers and their arrays are shown below.

1

4

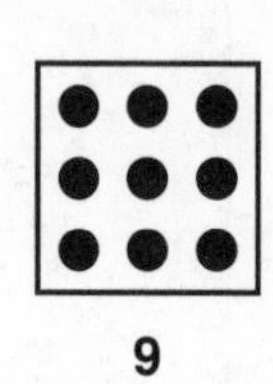
9

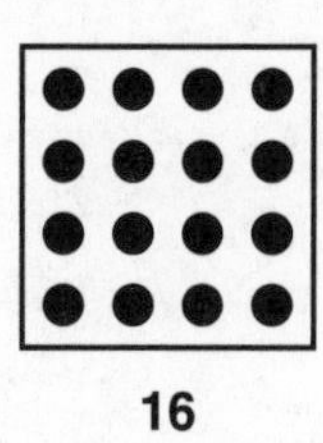
16

1. Draw a square array for the next square number after 16.

   Square number: ________

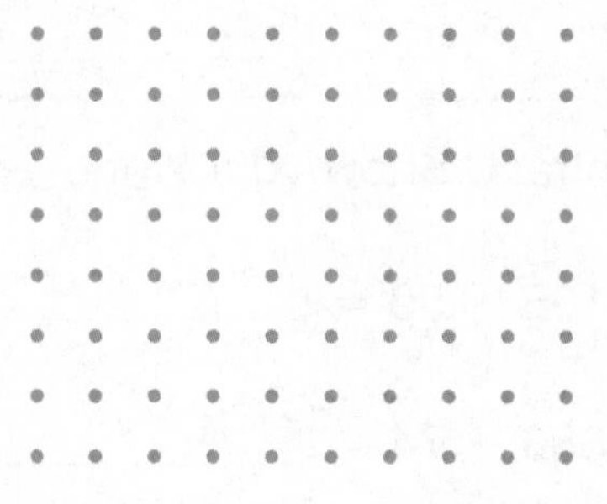

2. List all the square numbers through 100. Use counters or draw arrays, if you need help.

   ____

3. Can a square number be a prime number? ________ Why or why not?

   ____

   ____

4. Notice which square numbers are even and which are odd. Can you find a pattern? ________ If yes, describe the pattern.

   ____

   ____

Square Numbers (cont.)

Any square number can be written as the product of a number multiplied by itself.

Example The third square number, 9, can be written as $3 * 3$.

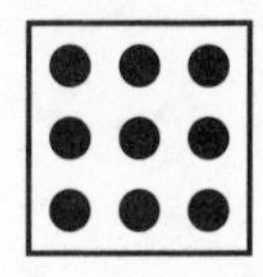

There is a shorthand way of writing square numbers: $9 = 3 * 3 = 3^2$.

You can read 3^2 as "3 times 3," "3 squared," or "3 to the second power." The raised 2 is called an **exponent.** It tells that 3 is used as a factor 2 times. Numbers written with an exponent are said to be in **exponential notation.**

Be careful! The number 3^2 is not the same as the product of $3 * 2$.
3^2 equals $3 * 3$, which is 9, and $3 * 2 = 6$.

5. Fill in the blanks.

Product	Exponential Notation	Square Number
$4 * 4$	4^2	16
$7 * 7$		
$10 * 10$		
____ * ____	11^2	

Some calculators have a key with the symbol [^] on it. It is called the **exponent key** and can be used to find the square of a number.

6. Press 3 [^] 2 [Enter]. What does the display show? ________

If your calculator has an exponent key, use it to solve the following problems. If not, you can use the multiplication key.

7. $8^2 =$ ________ **8.** $12^2 =$ ________ **9.** $14^2 =$ ________

10. $20^2 =$ ________ **11.** $43^2 =$ ________ **12.** $67^2 =$ ________

13. Start with 4. Square it. Now square the result. What do you get? ________

Date Time

Math Boxes 1.7

1. Complete.

a. 900 * 800 = ______________

b. 5,000 * ______________ = 300,000

c. 5,400 = ______________ * 60

d. 42,000 = ______________ * 700

e. 1,500 = ______________ * 3

2. a. How many crayons are there in 10 boxes, if each box contains 48 crayons?

______________ (unit)

b. How many crayons are there in 1,000 boxes?

______________ (unit)

3. Complete the table.

Fraction	Decimal	Percent
$\frac{1}{2}$		
	0.125	
		80%
$\frac{3}{4}$		
		32%

4. a. Write a 6-digit numeral with
4 in the hundredths place,
3 in the hundreds place,
6 in the thousands place,
5 in the tens place,
and 2s in all other places.

___, ___ ___ ___ . ___ ___

b. Write this numeral in words.

____

____

____

5. Circle the numbers that are divisible by 6.

438 629 702 320 843

6. Round 15,783,406 to the nearest …

a. million. ______________

b. thousand. ______________

c. hundred-thousand. ______________

Date Time

Unsquaring Numbers

You know that $6^2 = 6 * 6 = 36$. The number 36 is called the **square** of 6. If you **unsquare** 36, the result is 6. The number 6 is called the **square root** of 36.

1. "Unsquare" each number. The result is its square root. Do not use the [√] key on the calculator.

Example __12__ ^{2} = 144 The square root of 144 is __12__.

a. ________^{2} = 225 The square root of 225 is ________.

b. ________^{2} = 729 The square root of 729 is ________.

c. ________^{2} = 1,600 The square root of 1,600 is ________.

d. ________^{2} = 361 The square root of 361 is ________.

2. Which of the following are square numbers? Circle them.

576 794 1,044 4,356 6,400 5,770

Comparing Numbers with Their Squares

3. a. Unsquare the number 1. ________^{2} = 1

b. Unsquare the number 0. ________^{2} = 0

4. a. Is 5 greater than or less than 1? ____________________

b. 5^2 = ________ **c.** Is 5^2 greater than or less than 5? ____________________

5. a. Is 0.50 greater than or less than 1? ____________________

b. Use your calculator. 0.50^2 = ________

c. Is 0.50^2 greater than or less than 0.50? ____________________

6. a. When you square a number, is the result always greater than the number you started with? ________

b. Can it be less? ________

c. Can it be the same? ________

Date Time

Math Boxes 1.8

1. Write < or >.

   a. 3.8 ________ 0.83

   b. 0.4 ________ 0.30

   c. 6.24 ________ 6.08

   d. 0.05 ________ 0.5

   e. 7.12 ________ 7.2

2. Round each number to the nearest thousand.

   a. 8,692 ____________________

   b. 49,573 ____________________

   c. 2,601,458 ____________________

   d. 300,297 ____________________

   e. 599,999 ____________________

3. Subtract. Show your work.

   a. $\begin{array}{r} 777 \\ -\ 259 \\ \hline \end{array}$

   b. $\begin{array}{r} 508 \\ -\ 125 \\ \hline \end{array}$

   c. $\begin{array}{r} 5,009 \\ -\ 188 \\ \hline \end{array}$

   d. $\begin{array}{r} 8,435 \\ -\ 997 \\ \hline \end{array}$

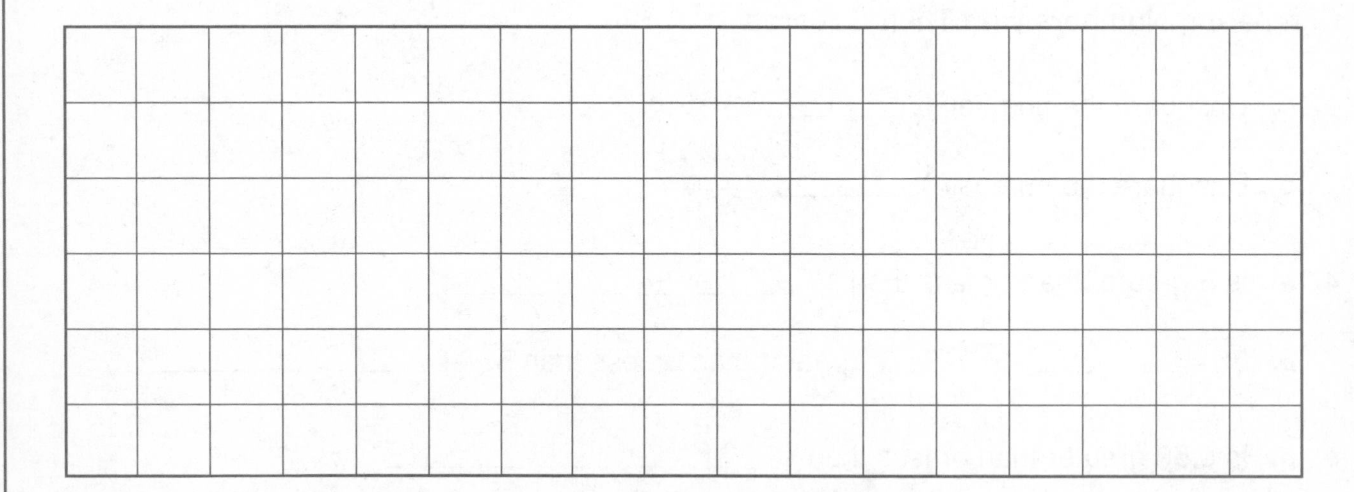

4. List all of the factors of 64.

   ____

5. In the morning, I need 30 minutes to shower and dress, 15 minutes to eat, and another 15 minutes to ride my bike to school. School begins at 8:30 A.M. What is the latest I can get up and still get to school on time?

   ____

Date Time

Factor Strings

A **factor string** is a name for a number written as a product of two or more factors. In a factor string, 1 may not be used as a factor.

The **length of a factor string** is equal to the number of factors in the string. The longest factor string for a number is made up of prime numbers. The longest factor string for a number is called the **prime factorization** of that number.

Example

Number	Factor Strings	Length
20	2 * 10	2
	4 * 5	2
	2 * 2 * 5	3

The order of the factors is not important. For example, 2 * 10 and 10 * 2 are the same factor string.

The longest factor string for 20 is 2 * 2 * 5. So the prime factorization of 20 is 2 * 2 * 5.

1. Find all the factor strings for each number below.

a.

Number	Factor Strings	Length
12		

b.

Number	Factor Strings	Length
16		

c.

Number	Factor Strings	Length
18		

d.

Number	Factor Strings	Length
28		

Date Time

Factor Strings (cont.)

2. Write the prime factorization (the longest factor string) for each number.

a. 27 = ____________________

b. 40 = ____________________

c. 36 = ____________________

d. 42 = ____________________

e. 48 = ____________________

f. 60 = ____________________

g. 100 = ____________________

An **exponent** is a raised number that shows how many times the number to its left is used as a factor.

Examples 5^2 ← exponent

5^2 means 5 * 5, which is 25.
5^2 is read as "5 squared" or as "5 to the second power."

10^3 ← exponent

10^3 means 10 * 10 * 10, which is 1,000.
10^3 is read as "10 cubed" or as "10 to the third power."

2^4 ← exponent

2^4 means 2 * 2 * 2 * 2, which is 16.
2^4 is read as "2 to the fourth power."

3. Write each number as a product of factors. Then find the answer.

Examples 2^3 = 2 * 2 * 2 = 8

$2^2 * 9$ = 2 * 2 * 9 = 36

a. 10^4 = ____________________ = __________

b. $3^2 * 5$ = ____________________ = __________

c. $2^4 * 10^2$ = ____________________ = __________

4. Rewrite each product using exponents.

Examples 5 * 5 * 5 = 5^3 5 * 5 * 3 * 3 = $5^2 * 3^2$

a. 3 * 3 * 3 * 3 = __________

b. 4 * 7 * 7 = __________

c. 2 * 5 * 5 * 7 = __________

d. 2 * 2 * 2 * 5 * 5 = __________

Date Time

Review

1. Circle the square numbers.

10 16 24 64 81 48

2. List the factors of each number from least to greatest.

a. 15 ______

b. 28 ______

c. 30 ______

d. 36 ______

3. Do not use a calculator to solve the problems.
Circle the numbers that are:

a. divisible by 2	3,336	5,027	19,008
b. divisible by 3	1,752	497	28,605
c. divisible by 5	2,065	12,340	10,003
d. divisible by 9	921	5,715	36,360

4. Circle the prime numbers.

7 14 1 25 39 41

5. Write the prime factorization for each number.

a. 12 ______

b. 20 ______

c. 49 ______

d. 32 ______

6. Fill in the missing numbers.

a. $5^2 =$ ______

b. ______$^2 = 36$

c. $1^2 + 2^2 + 3^2 =$ ______

Date Time

Math Boxes 1.9

1. Complete.

   a. 300 * 40 = ____________

   b. ____________ = 80 * 200

   c. ____________ = 900 * 600

   d. 6,400 = ____________ * 80

   e. 36,000 = 600 * ____________

2. a. How many marbles are there in 7 bags, if each bag contains 8 marbles?

   ____________ (unit)

   b. How many marbles are there in 700 bags?

   ____________ (unit)

3. Complete the table.

Fraction	Decimal	Percent
$\frac{3}{8}$		
		60%
$\frac{2}{5}$		
	0.55	
$\frac{8}{100}$		

4. a. Write a 6-digit numeral with 7 in the thousands place, 5 in the hundredths place, 4 in the tenths place, 3 in the tens place, and 9s in all other places.

   ___, ___ ___ ___ . ___ ___

   b. Write this numeral in words.

   ____

   ____

   ____

   ____

5. Circle the numbers that are divisible by 9.

   360 252 819 426 651

6. Round 385.27 to the nearest …

   a. hundred. ____________

   b. whole number. ____________

   c. tenth. ____________

Date Time

Time to Reflect

1. Describe what you liked or did not like about playing the game *Factor Captor.*

2. Explain how making an array might help someone find factors of a number.

Look back through journal pages 2–24.

3. What activity or lesson did you enjoy most in this unit and what did you learn from it?

4. What was your least favorite lesson or activity in this unit and why?

Date Time

Math Boxes 1.10

1. a. Write a 7-digit numeral with
3 in the tens place,
5 in the hundredths place,
7 in the hundreds place,
2 in the ten-thousands place,
and 4s in all other places.

___ ___, ___ ___ ___ . ___ ___

b. Write this numeral in words.

____

____

____

2. Phoebe received these math test scores:
93, 96, 85, 100, 98, 100, 99, 95.

a. Maximum ___________

b. Minimum ___________

c. Range ___________

d. Median ___________

3. Complete.

a. 27,000 = ___________ * 90

b. ___________ = 800 * 600

c. ___________ = 700 * 8,000

d. ___________ = 50 * 600

e. 350 = 7 * ___________

4. Write < or >.

a. 0.90 ________ 0.89

b. 3.52 ________ 3.8

c. 6.91 ________ 6.3

d. 4.05 ________ 4.2

e. 0.38 ________ 0.5

5. Solve.

a. $\begin{array}{r} 207 \\ -\ 158 \\ \hline \end{array}$

b. $\begin{array}{r} 325 \\ +\ 116 \\ \hline \end{array}$

c. $\begin{array}{r} 829 \\ +\ 580 \\ \hline \end{array}$

d. $\begin{array}{r} 628 \\ -\ 444 \\ \hline \end{array}$

e. $\begin{array}{r} 385 \\ -\ 179 \\ \hline \end{array}$

f. $\begin{array}{r} 523 \\ +\ 478 \\ \hline \end{array}$

Date Time

Estimation Challenge

Sometimes you will be asked to solve a problem for which it is difficult, or even impossible, to find an **exact** answer. Your job will be to make your best estimate and then defend it. We call this kind of problem an **Estimation Challenge.**

Estimation Challenges can be difficult and they take time to solve. Usually, you will work with a partner or as part of a small group.

Estimation Challenge Problem

Imagine that you are living in a time when there are no cars, trains, or planes. You do not own a horse, a boat, or any other means of transportation.

You plan to travel to ______________________. You will have to walk there.
(location given by your teacher)

Information needed to solve the problem.

Definition of a "step." 1 mile = 5,280 feet

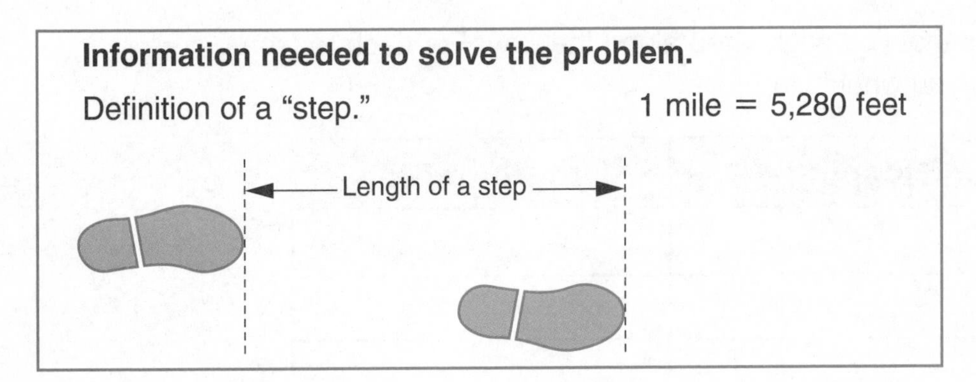

1. About how many miles is it from your school to your destination?

   About ______________________ miles

2. a. About how many footsteps will you have to take to get from your school to your destination?

   About ______________________ footsteps

   b. What did you do to estimate the number of footsteps you would take?

   ____

   ____

   ____

Date Time

Estimation Challenge (cont.)

3. a. Suppose that you did not stop to rest, eat, sleep, or for any other reason. About how long would it take you to get from school to your destination?

   About ______________________ hours

   b. What did you do to estimate how long it would take you?

   ____

   ____

   ____

4. Suppose you start from school at 7:00 A.M. on Monday. You take time out to rest, eat, sleep, and for other reasons.

   a. List all of the reasons that you might stop along the way. For each reason, write about how long you would stop.

Reason for Stopping	Length of Stop

   b. At about what time, and on what day of the week, would you expect to reach your destination?

   Time: About ______________________ Day: ______________________

5. Who did you work with on this Estimation Challenge? ______________________

Date Time

Math Boxes 2.1

1. Find the missing numbers and landmarks for the set of numbers:

18, 20, 20, 24, 27, 27, ________, 30, 33, 34, 36, 36, ________

a. Range: 22

b. Mode: 27

c. Minimum: ________

d. Maximum: ________

2. Sam drew a trapezoid and a square and covered them as shown. Write the name below each figure. Then finish each drawing.

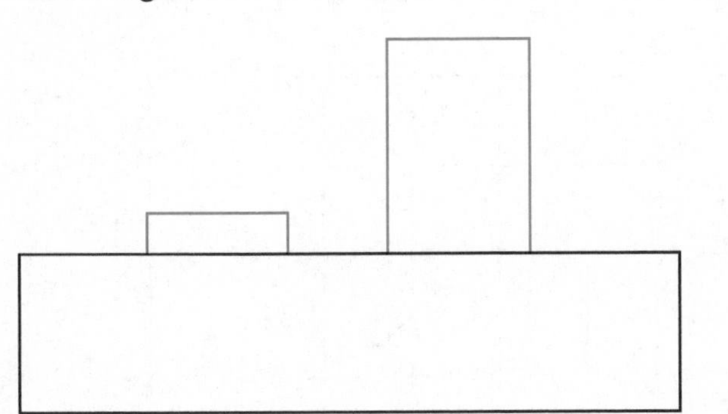

SRB 136

3. Write the following numbers in standard notation.

a. 3^3 = ________

b. 7^2 = ________

c. 4^3 = ________

d. 5^2 = ________

e. 2^4 = ________

4. a. How many dots are in this array?

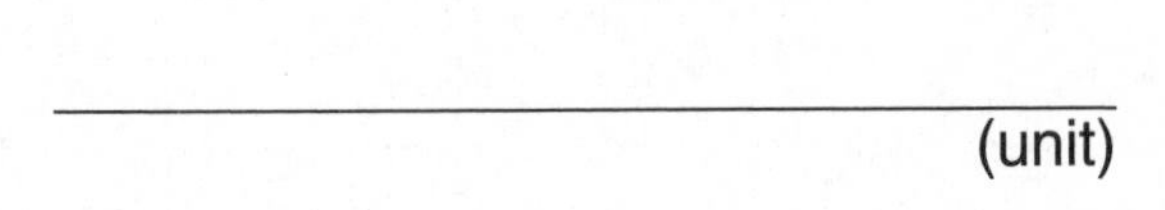

________ (unit)

b. Write a number model for the array.

____

5. a. Build an 8-digit numeral. Write
7 in the ten-millions place,
2 in the tens place,
4 in the hundred-thousands place,
6 in the ones place,
and 5 in all the other places.

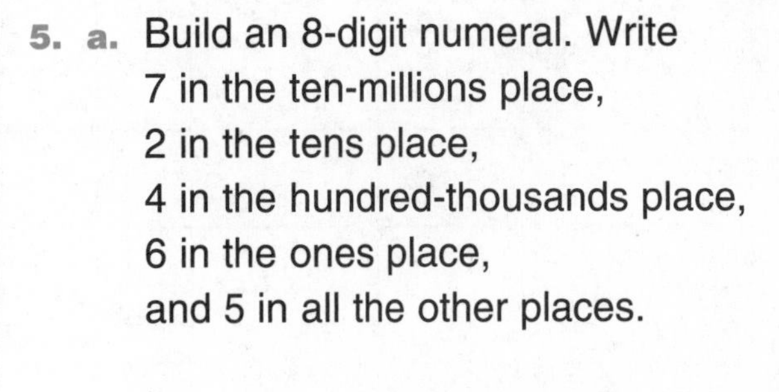

___ ___, ___ ___ ___, ___ ___ ___

b. Write this numeral in words.

____

____

____

Date Time

Methods for Addition

Solve Problems 1 and 2 using the partial-sums method. Solve Problems 3 and 4 using the column-addition method. Solve the rest of the problems using any method you choose. Show your work in the space below. Compare your answers with your partner's answers. Resolve any disagreements.

1. 714 + 468 = ______________
2. 253 + 187 = ______________
3. ______________ = 45.6 + 17.3
4. 475 + 39 + 115 + 65 = ______________
5. 234.1 + 27.6 = ______________
6. ______________ = 217 + 192 + 309
7. 3,416 + 2,795 = ______________
8. ______________ = 36.47 + 9.58

Date Time

Math Boxes 2.2

1. Measure $\angle TAG$ to the nearest degree.

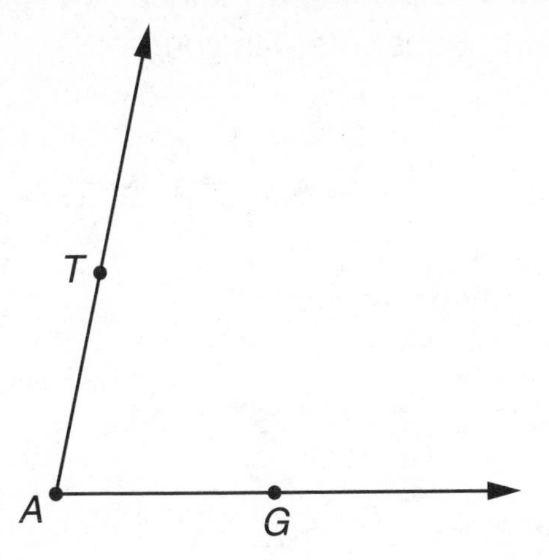

$\angle TAG$: ________

2. Write < or >.

a. 0.17 ________ 1.7

b. 5.4 ________ 5.04

c. 0.03 ________ 0.1

d. 2.24 ________ 2.2

e. 1.9 ________ 1.89

3. Write the prime factorization of 72.

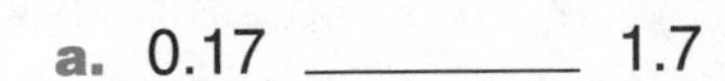

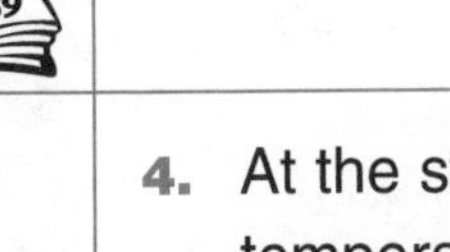

4. At the start of an experiment, the temperature in a box was 27°C. The temperature was increased by 32 degrees. Next it was decreased by 43 degrees. What was the temperature in the box then?

____

5. Tell whether the following numbers are prime or composite.

a. Number of feet in $\frac{2}{3}$ yard ________

b. Number of seconds in $\frac{1}{2}$ minute ________

c. Number of millimeters in 3.3 centimeters ________

d. Number of hours in $\frac{1}{8}$ day ________

e. Number of inches in $\frac{1}{6}$ yard ________

Date Time

Methods for Subtraction

Solve Problems 1 and 2 using the trade-first method. Solve Problems 3 and 4 using the partial-differences method. Solve the rest of the problems using any method you choose. Show your work in the space below. Compare your answers with your partner's answers. Resolve any disagreements.

1. 67 − 39 = ______________
2. ______________ = $34.68 − $15.75
3. 895 − 327 = ______________
4. 7,053 − 2,690 = ______________
5. 146.9 − 92.5 = ______________
6. ______________ = 138.2 − 79.6
7. ______________ = 5,829 − 673
8. 9.6 − 4.87 = ______________

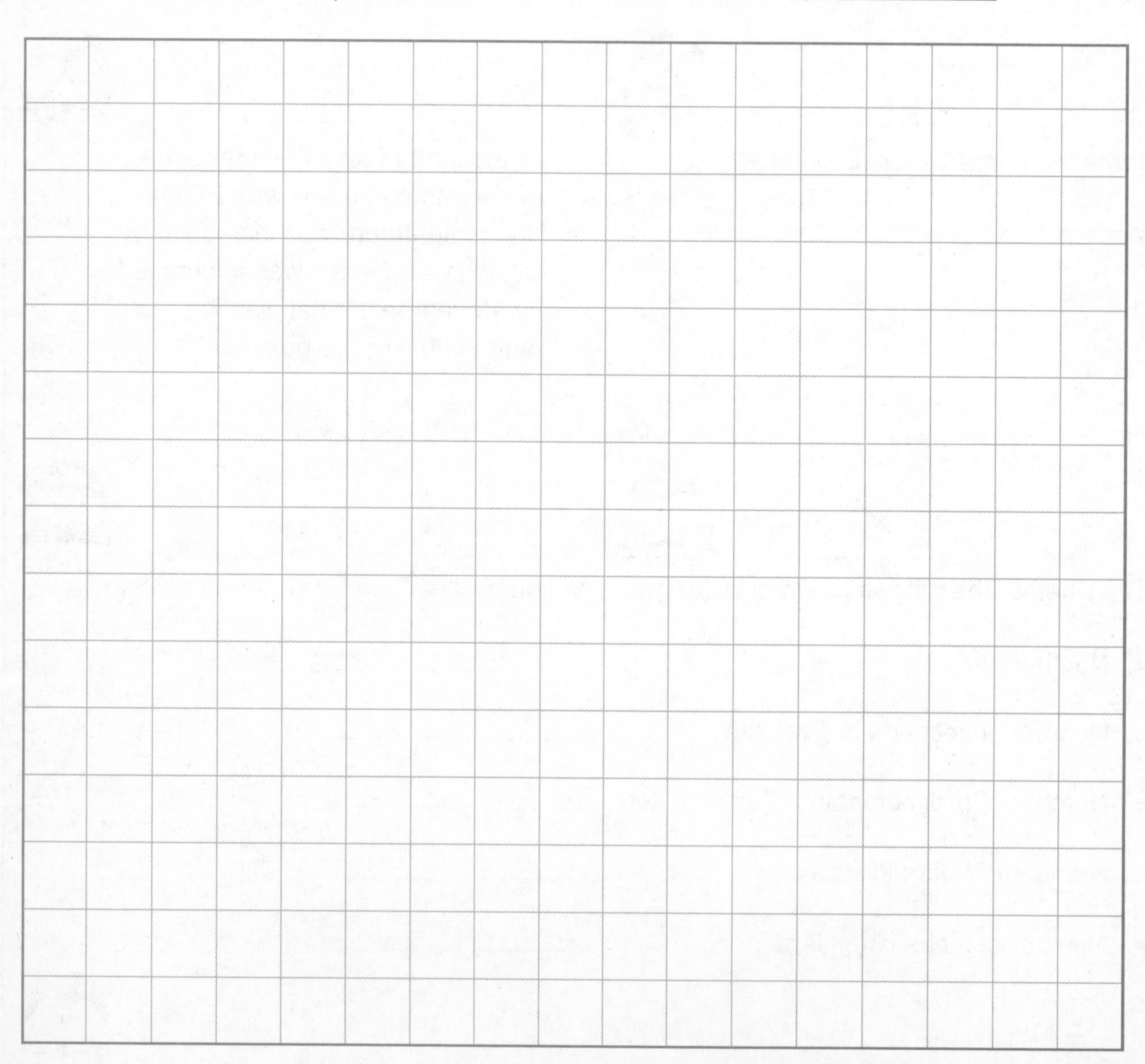

Date Time

Math Boxes 2.3

1. Find the missing numbers and landmarks for the set of numbers:

   48, 50, 51, 51, 57, 59, 60, 63, 69, ________, 76, ________

   a. Range: ________
   b. Mode: 76
   c. Minimum: ________
   d. Maximum: 76

2. I am a polygon. I have five sides and two right angles.

   a. Draw me in the space below.

   b. I am called a ____________________.

3. Write the following numbers in standard notation.

   a. 6^2 = ____________
   b. 10^5 = ____________
   c. 14^2 = ____________
   d. 8^3 = ____________
   e. 3^4 = ____________

4. a. How many dots are in this array?

   ______________________________ (unit)

   b. Write a number model for the array.

   ____

5. a. Build a 9-digit numeral. Write
   2 in the hundreds place,
   5 in the ten-thousands place,
   7 in the millions place,
   6 in the hundred-millions place,
   and 3 in all other places.

   __ __ __ , __ __ __ , __ __ __

   b. Write this numeral in words.

   ____
   ____
   ____
   ____

Date Time

Addition and Subtraction Number Stories

For each problem on pages 36 and 37, fill in the blanks and solve the problem.

Example Maria had 2 decks of cards. One of the decks had only 36 cards instead of 52. The rest were missing from the deck. How many cards were missing?

- List the numbers needed to solve the problem. 36 and 52
- Describe what you want to find. The number of missing cards
- Write an open sentence: $36 + c = 52$
- Find the missing number in the open sentence. Solution: 16
- Write the answer to the number story. Answer: 16 cards (unit)

1. Anthony got a new bike. He rode 18 miles the first week, 27 miles the second week, and 34 miles the third week. How many miles did he ride altogether?

a. List the numbers needed to solve the problem. ______________

b. Describe what you want to find. ______________

____

c. Open sentence: ______________

d. Solution: ______________ **e.** Answer: ______________ (unit)

2. Regina has \$23.08. Her sister has \$16.47. Her brother has only \$5.00. How much more money does Regina have than her sister?

a. List the numbers needed to solve the problem. ______________

b. Describe what you want to find. ______________

____

c. Open sentence: ______________

d. Solution: ______________ **e.** Answer: ______________

3. Lucas was having 12 friends over for breakfast. He started with 19 eggs. He bought 1 dozen more eggs. How many eggs did he have to cook for breakfast?

a. List the numbers needed to solve the problem. ______________

b. Describe what you want to find. ______________

____

c. Open sentence: ______________

d. Solution: ______________ **e.** Answer: ______________ (unit)

Date Time

Addition and Subtraction Number Stories (cont.)

4. Nicholas earned $48 mowing lawns one weekend. With the money he earned, he bought 2 CDs that cost a total of $23. How much money did he have left?

   a. List the numbers needed to solve the problem. ______

   b. Describe what you want to find. ______

   ______

   c. Open sentence: ______

   d. Solution: ______ e. Answer: ______

Circle the open sentence that best matches each story and then solve the problem.

5. Patrick's hobby is to paint color-by-number pictures. He spent 24 hours painting in June and 37 hours painting in July. The last picture he painted had 18 different colors. How many hours did he paint in the two months?

   $18 + h = 37$ $24 + h = 37$

   $37 + 24 = h$ $37 - h = 18$

   Answer: ______ (unit)

6. Sue walked 2 miles to Jan's house. Then both girls walked 2 miles to Tad's house. Sue took 28 minutes to get to Jan's house. The girls took 45 minutes to get to Tad's house. How much longer did it take to get to Tad's house than to Jan's house?

   $2 * 28 = m$ $2 + 28 + m = 45$

   $m - 28 = 45$ $45 - 28 = m$

   Answer: ______ (unit)

7. Write and solve your own number story. ______

   ______

   ______

   a. List the numbers needed to solve the problem. ______

   b. Describe what you want to find. ______

   ______

   c. Open sentence: ______

   d. Solution: ______ e. Answer: ______ (unit)

Date Time

Math Boxes 2.4

1. Measure $\angle BOP$ to the nearest degree.

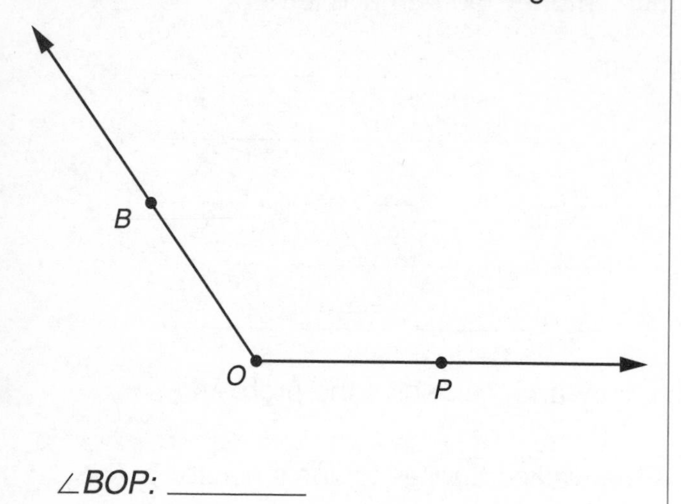

$\angle BOP$: ________

2. Write < or >.

a. 3.67 ________ 3.7

b. 0.02 ________ 0.21

c. 4.06 ________ 4.02

d. 3.1 ________ 3.15

e. 7.6 ________ 7.56

3. Write the prime factorization of 32.

____

4. The temperature at midnight was 25°F. The windchill temperature was 14°F. How much warmer was the actual temperature than the windchill temperature?

____

5. Tell whether the following numbers are prime or composite.

a. The number of millimeters in 1.7 cm ________

b. The number of degrees in a right angle ________

c. The number of inches in $\frac{11}{4}$ feet ________

d. One less than the number of hours in 1 day ________

e. The number of months in $\frac{1}{4}$ of a year ________

Date Time

Estimating Your Reaction Time

Tear out Activity Sheet 3 from the back of your journal. Cut out the Grab-It Gauge.

It takes two people to perform this experiment. The *tester* holds the Grab-It Gauge at the top. The *contestant* gets ready to catch the gauge by placing his or her thumb and index finger at the bottom of the gauge, *without quite touching it.* (See the picture.)

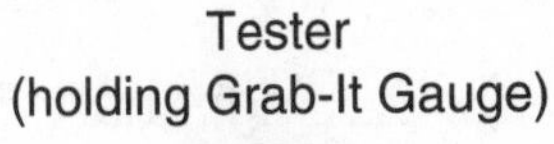

Tester
(holding Grab-It Gauge)

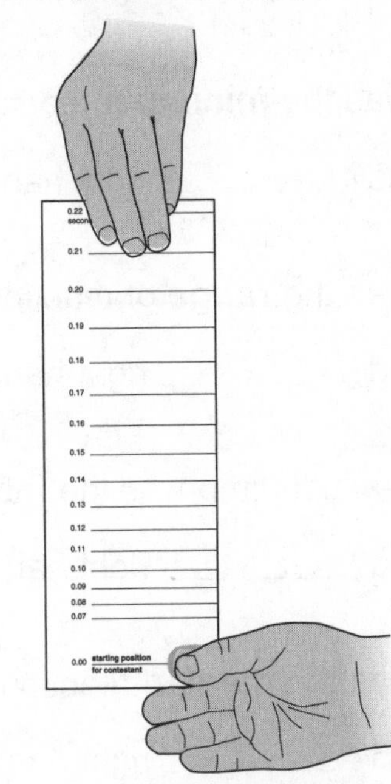

Contestant
(not quite touching Grab-It Gauge)

When the contestant is ready, the tester lets go of the gauge. The contestant tries to grab it with his or her thumb and index finger as quickly as possible.

The number grasped by the contestant shows that person's reaction time, to the nearest hundredth of a second. The contestant then records that reaction time in the data table shown below.

Partners take turns being tester and contestant. Each person should perform the experiment 10 times with each hand.

Reaction Time (in seconds)			
Left Hand		**Right Hand**	
1.	6.	1.	6.
2.	7.	2.	7.
3.	8.	3.	8.
4.	9.	4.	9.
5.	10.	5.	10.

Date Time

Estimating Your Reaction Time (cont.)

Use the results of your Grab-It experiment to answer the following questions.

1. What was the **maximum** reaction time for your

   left hand? ________ right hand? ________

2. What was the **minimum** reaction time for your

   left hand? ________ right hand? ________

3. What was the **range** of reaction times for your

   left hand? ________ right hand? ________

4. What reaction time was the **mode** for your

   left hand? ________ right hand? ________

5. What was the **median** reaction time for your

   left hand? ________ right hand? ________

6. What was the **mean** reaction time for your

   left hand? ________ right hand? ________

7. If you could use just one number to estimate your reaction time, which number would you choose as the best estimate? Circle one.

   minimum maximum mode median mean

   Explain. ____

   ____

8. Which of your hands reacted more quickly in the Grab-It experiment?

   ____

Date Time

Driving Decimals

The Indianapolis 500 is a car race held each year at the Indianapolis Motor Speedway. The racers drive more than 200 laps on a $2\frac{1}{2}$-mile oval track.

The table at the right shows the 10 fastest winning speeds from various years for this race. Use the table to answer each question below.

Fastest Winning Speeds for the Indianapolis 500

Driver	Year	Speed (mph)
Arie Luyendyk	1990	185.981
Rick Mears	1991	176.457
Bobby Rahal	1986	170.722
Emerson Fittipaldi	1989	167.581
Rick Mears	1984	163.612
Mark Donohue	1972	162.962
Al Unser	1987	162.175
Tom Sneva	1983	162.117
Gordon Johncock	1982	162.029
Al Unser	1978	161.363

Source: The World Almanac and Book of Facts 2000

1. a. What was Emerson Fittipaldi's winning speed for the Indianapolis 500?

   ______________________ (unit)

   b. In what year did he set this speed record?

   ____

2. How much faster was Rick Mears's speed in 1991 than in 1984?

   ______________________ (unit)

3. What is the range of speeds in the table? ______________________ (unit)

   Reminder: The range is the difference between the fastest speed and the slowest speed.

4. a. Which two drivers have the smallest difference between their winning speeds?

   ____

   b. What is the difference between the two speeds? ______________________ (unit)

Challenge

5. What is the median of the speeds in the table? ______________________ (unit)

Date Time

Math Boxes 2.5

1. I have four sides. All opposite sides are parallel. I have no right angles.

a. Draw me in the space below.

b. I am called a ______________.

2. Write < or >.

a. 0.45 ______ $\frac{3}{4}$

b. 0.89 ______ $\frac{8}{10}$

c. $\frac{4}{5}$ ______ 0.54

d. $\frac{1}{3}$ ______ 0.35

e. $\frac{7}{8}$ ______ 0.9

3. Complete each pattern.

a. 25, ______, 61, ______

b. 87, ______, 43, ______

c. 21, ______, 49, ______

d. 64, ______, ______, ______, 32

e. 61, ______, ______, ______, 81

4. Solve.

	Solution
a. $23 + x = 60$	x = ______
b. $36 = p * 4$	p = ______
c. $200 = 50 * m$	m = ______
d. $55 + t = 70$	t = ______
e. $28 - b = 13$	b = ______

5. Add.

a. 632 + 859

b. 2.24 + 3.85

c. 1,902 + 478

d. 3,341 + 799

e. 1,654 + 2,020

Date Time

Describing Chances

1. Draw a line from each spinner to the number that best describes the chance of landing in the blue area.

Spinner	Chance of Landing on Blue
	0.25
	50%
	$\frac{2}{3}$
	0.75
	90%

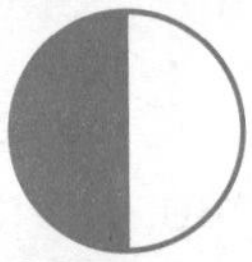

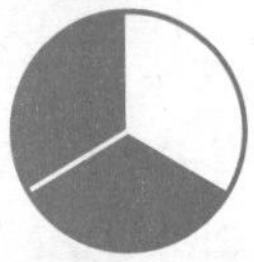

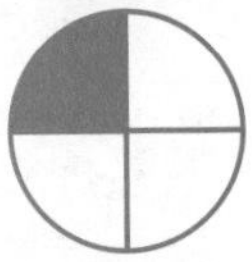

2. Draw a line from each event listed below to the best description of the chance that the event will happen.

Example Most people will fly in an airplane at least once during their lifetime. Therefore, draw a line to "extremely likely."

Event

a. A person will fly in an airplane at least once during his or her lifetime.

b. The sun will rise tomorrow.

c. An adult is able to swim.

d. A newborn baby will be a girl.

e. A long-distance call will result in a busy signal.

f. There will be an earthquake in California during the next year.

g. Your home will catch on fire during the next year.

Chance

certain

extremely likely

very likely

likely

50–50 chance

unlikely

very unlikely

extremely unlikely

impossible

Date Time

A Thumbtack Experiment

Make a guess: If you drop a thumbtack, is it more likely

to land with the point up or with the point down? ______________________

The experiment described below will enable you to make a careful estimate of the chance that a thumbtack will land point down.

1. Work with a partner. You should have 10 thumbtacks and 1 small cup. Do the experiment at your desk or table so that you are working over a smooth, hard surface.

Place the 10 thumbtacks inside the cup. Shake the cup a few times, and then drop the tacks on the desk surface. Record the number of thumbtacks that land point up and the number that land point down.

Toss the 10 thumbtacks 9 more times and record the results each time.

Toss	Number Landing Point Up	Number Landing Point Down
1		
2		
3		
4		
5		
6		
7		
8		
9		
10		
	Total Up =	**Total Down =**

2. In making your 10 tosses, you dropped a total of 100 thumbtacks.

What fraction of the thumbtacks landed point down? ________

3. Write this fraction on a small stick-on note. Also, write it as a decimal and as a percent.

4. *Do this later:* For the whole class, the chance a tack lands point down is ________ .

Date Time

Describing a Set of Data

1. Justin, Vincent, Gregory, Bernard, Melinda, Frieda, and Marina estimated the number of jellybeans in a jar. They made the following estimates:

Justin	247
Vincent	375
Gregory	199
Bernard	252
Melinda	305
Frieda	200
Marina	299

a. What was the minimum estimate? ________

b. What was the maximum estimate? ________

c. What was the mode of the estimates? ________

d. What was the range of the estimates? ________

e. What was the median estimate? ________

f. There were 270 jellybeans in the jar. Whose estimate was closest? ________

2. Eight friends were comparing their science test scores. There were 50 questions on the test. Their scores were as follows:

80, 96, 88, 100, 88, 94, 90, 88

a. What was the minimum score? ________

b. What was the maximum score? ________

c. What was the mode of the scores? ________

d. What was the range of the scores? ________

e. What was the median score? ________

f. Explain how you would find the mean for the eight scores. ________

____

____

Date Time

Math Boxes 2.6

1. Cross out the shapes below that are NOT polygons.

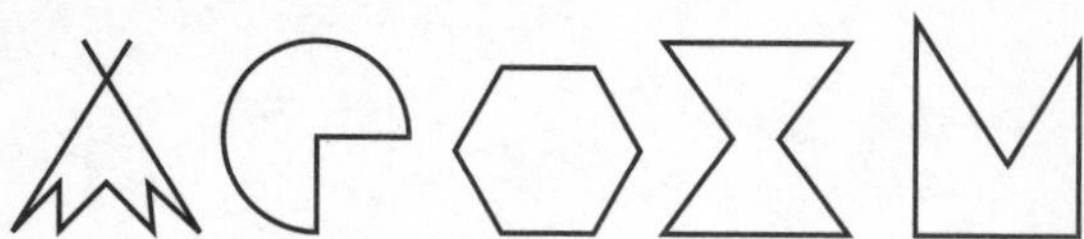

2. Find the perimeter of the rectangle.

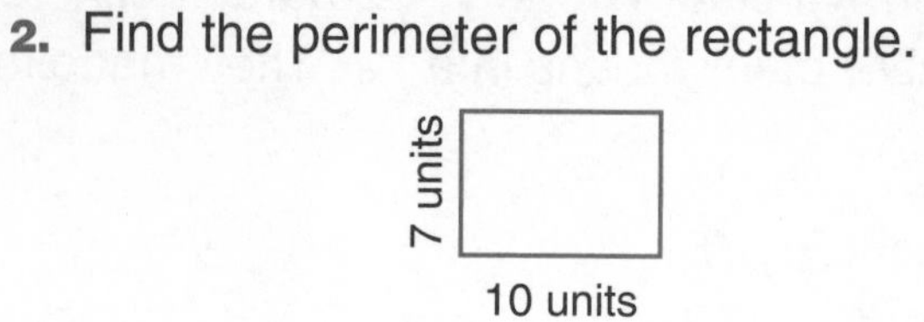

______________ (unit)

3. Multiply. Show your work.

a. 426 × 8

b. 395 × 26

c. 406 × 18

d. 297 × 53

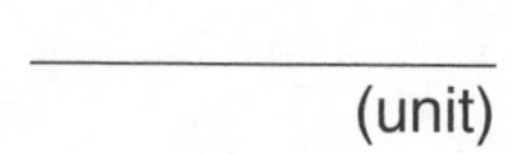

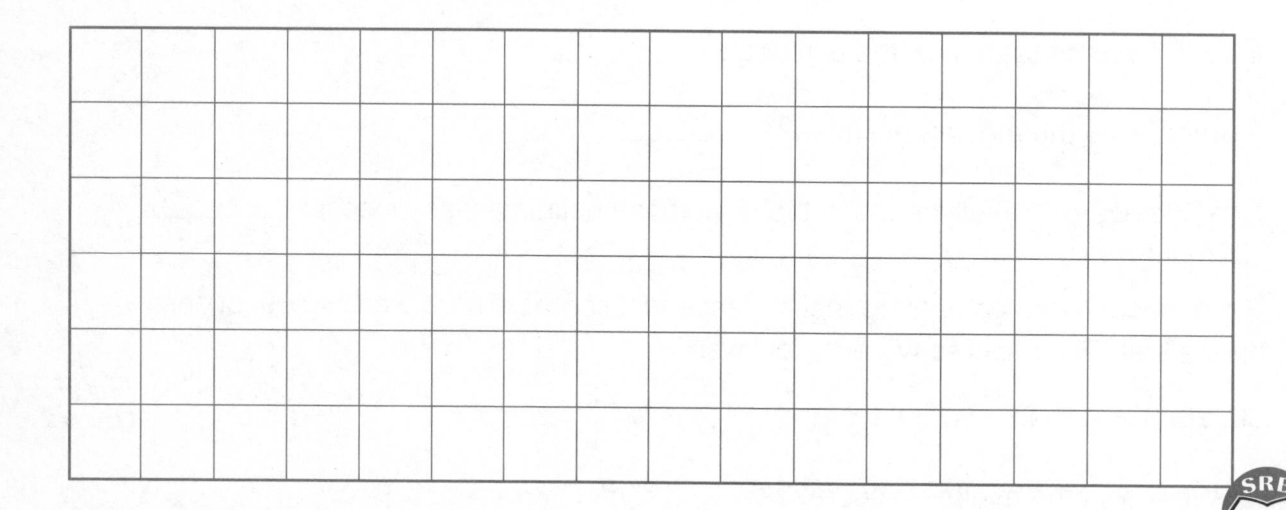

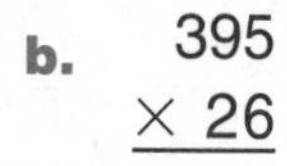

4. Give the value of the **boldface digit** in each numeral.

a. 2**8**7,051 ______________

b. 7,04**2**,690 ______________

c. 2**8**,609,381 ______________

d. 506,344,**5**26 ______________

e. **4**7,381,296 ______________

Date Time

Magnitude Estimates for Products

A **magnitude estimate** is a very rough estimate of the answer to a problem. A magnitude estimate will tell you if the exact answer is in the tenths, ones, tens, hundreds, thousands, and so on.

For each problem, make a magnitude estimate. Ask yourself, "Is the answer in the tenths, ones, tens, hundreds, thousands, or ten-thousands?" Circle the appropriate box. Do not solve the problems.

Example 14 * 17

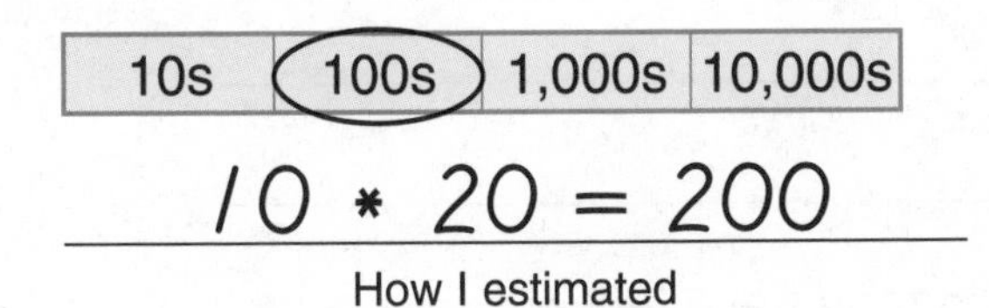

10s
100s
1,000s
10,000s
10 * 20 = 200

How I estimated

1. 56 * 37

10s	100s	1,000s	10,000s

How I estimated

2. 7 * 326

10s	100s	1,000s	10,000s

How I estimated

3. 95 * 48

10s	100s	1,000s	10,000s

How I estimated

4. 5 * 4,127

10s	100s	1,000s	10,000s

How I estimated

5. 46 * 414

10s	100s	1,000s	10,000s

How I estimated

6. 4.5 * 0.6

0.1s	1s	10s	100s

How I estimated

7. 7.6 * 9.1

0.1s	1s	10s	100s

How I estimated

8. 160 * 2.9

0.1s	1s	10s	100s

How I estimated

9. 0.8 * 0.8

0.1s	1s	10s	100s

How I estimated

Date Time

Solving Number Stories

For each problem, fill in the blanks and solve the problem.

1. Linell and Ben pooled their money to buy a video game. Linell had $12.40 and Ben had $15.88. How much money did they have in all?

   a. List the numbers needed to solve the problem. ____________________

   b. Describe what you want to find. ______________________________

   ____

   c. Open sentence: ______________________________________

   d. Solution: __________ e. Answer: ____________________

2. If the video game cost $22.65, how much money did they have left?

   a. List the numbers needed to solve the problem. ____________________

   b. Describe what you want to find. ______________________________

   ____

   c. Open sentence: ______________________________________

   d. Solution: __________ e. Answer: ____________________

3. Linell and Ben borrowed money so they could also buy a CD for $13.79. How much did they have to borrow so that they would have enough money to buy the CD?

   a. List the numbers needed to solve the problem. ____________________

   b. Describe what you want to find. ______________________________

   ____

   c. Open sentence: ______________________________________

   d. Solution: __________ e. Answer: ____________________

4. How much more did the video game cost than the CD?

   a. List the numbers needed to solve the problem. ____________________

   b. Describe what you want to find out. ____________________________

   ____

   c. Open sentence: ______________________________________

   d. Solution: __________ e. Answer: ____________________

Use with Lesson 2.7.

Date Time

Math Boxes 2.7

1. Look around the room and find an example of each of the following:

   a. a parallelogram ______

   b. a square ______

   c. a circle ______

   d. a polygon with more than 4 sides ______

   e. a cube ______

2. Subtract. Do not use a calculator.

   a. 1,924 − 385

   b. 7,431 − 5,555

   c. 1,493 − 208

   d. 322 − 199

   e. 602 − 483

3. Use a calculator to rename each of the following in standard notation.

   a. 24^2 = ______

   b. 11^3 = ______

   c. 9^4 = ______

   d. 4^5 = ______

   e. 2^7 = ______

4. Write five names for the number 23.

   a. ______

   b. ______

   c. ______

   d. ______

   e. ______

Date Time

Multiplication of Whole Numbers

For each problem, make a magnitude estimate. Circle the appropriate box. Do not solve the problems.

1. 6 * 543

10s	100s	1,000s	10,000s

How I estimated

2. 3 * 284

10s	100s	1,000s	10,000s

How I estimated

3. 46 * 97

10s	100s	1,000s	10,000s

How I estimated

4. 4 * 204

10s	100s	1,000s	10,000s

How I estimated

5. 25 * 37

10s	100s	1,000s	10,000s

How I estimated

6. 56 * 409

10s	100s	1,000s	10,000s

How I estimated

7. Solve each problem above for which your estimate is at least 1,000. Use the partial-products method for at least one problem. Show your work on the grid below.

Date Time

Multiplication of Decimals

For each problem, make a magnitude estimate. Circle the appropriate box. Do not solve the problems.

1. 2.4 * 63

0.1s	1s	10s	100s

How I estimated

2. 7.2 * 0.6

0.1s	1s	10s	100s

How I estimated

3. 13.4 * 0.3

0.1s	1s	10s	100s

How I estimated

4. 3.58 * 2.1

0.1s	1s	10s	100s

How I estimated

5. 7.84 * 6.05

0.1s	1s	10s	100s

How I estimated

6. 2.8 * 93.6

0.1s	1s	10s	100s

How I estimated

7. Solve each problem above for which your estimate is at least 10. Show your work on the grid below.

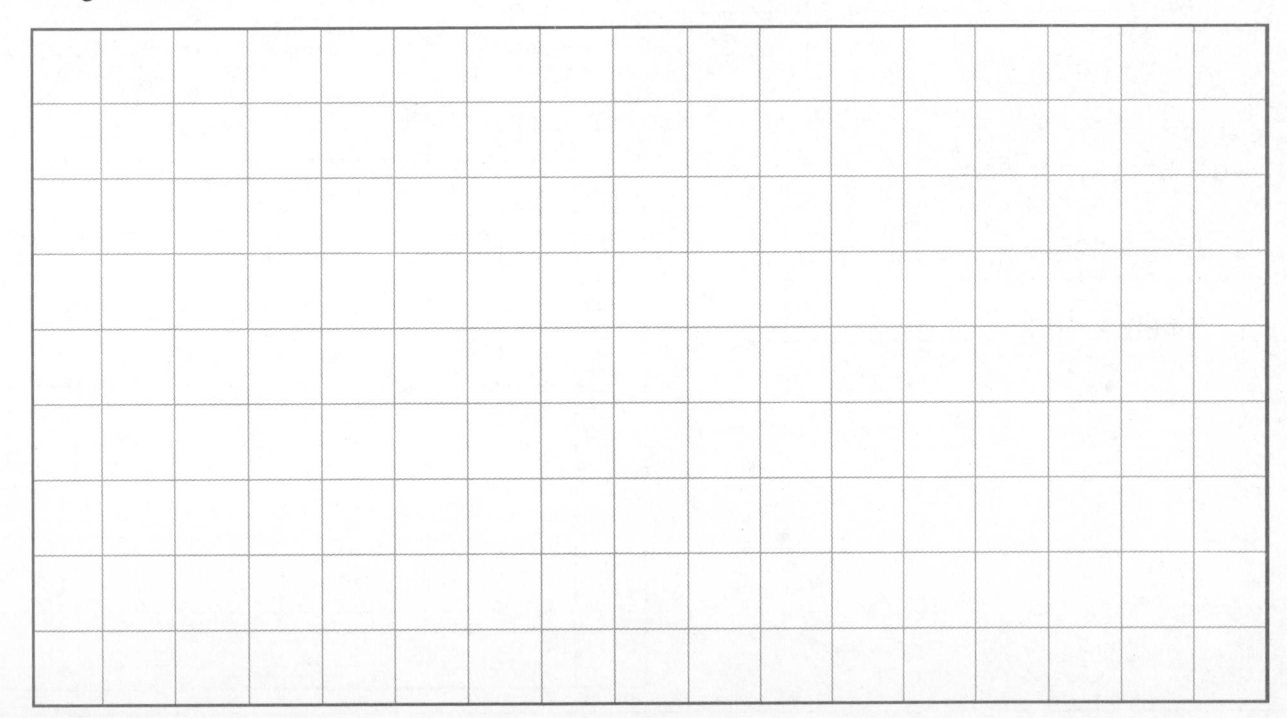

Date Time

Math Boxes 2.8

1. I have exactly six angles and all of my sides are the same length.

   a. Draw me in the space below.

   b. What shape am I?

   ____

2. Write < or >.

   a. $\frac{3}{5}$ ________ 0.70

   b. $\frac{1}{4}$ ________ 0.21

   c. 0.38 ________ $\frac{3}{10}$

   d. 0.6 ________ $\frac{2}{3}$

   e. 0.95 ________ $\frac{90}{100}$

3. Complete each pattern.

   a. 17, ________, ________, 62, ________

   b. 68, ________, ________, ________, 20

   c. 39, ________, ________, ________, 75

   d. 57, ________, ________, 33, ________

   e. 15, ________, ________, 33, ________

4. Solve.

	Solution
a. $5 * m = 45$	m = ________
b. $8 = 64 \div d$	d = ________
c. $8 = 48 \div k$	k = ________
d. $40 * s = 280$	s = ________
e. $w * 900 = 54{,}000$	w = ________

5. Add. Show your work.

   a. 885 + 329 = ____________

   b. 14.38 + 55.7 = ____________

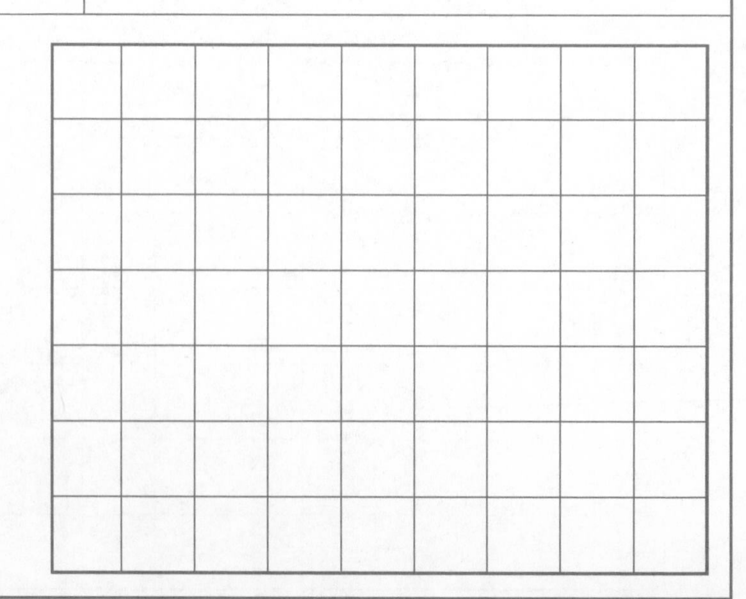

Date Time

Multiplication by the Lattice Method

For each problem:
- Make a magnitude estimate. Circle the appropriate box.
- Solve the problem using the lattice method. Show your work below.

1. 7 * 349 = ____________

10s	100s	1,000s	10,000s

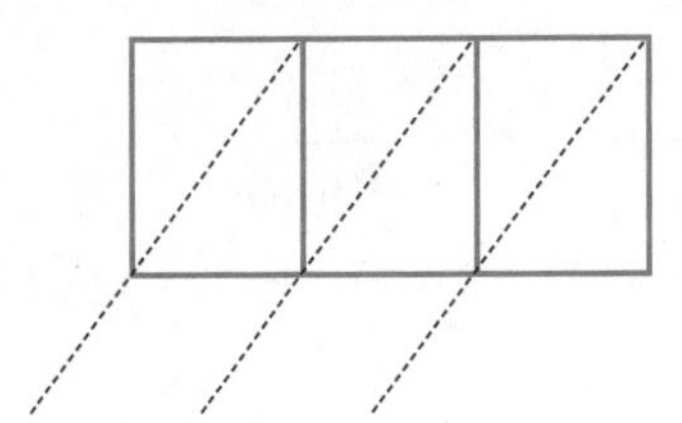

2. 48 * 72 = ____________

10s	100s	1,000s	10,000s

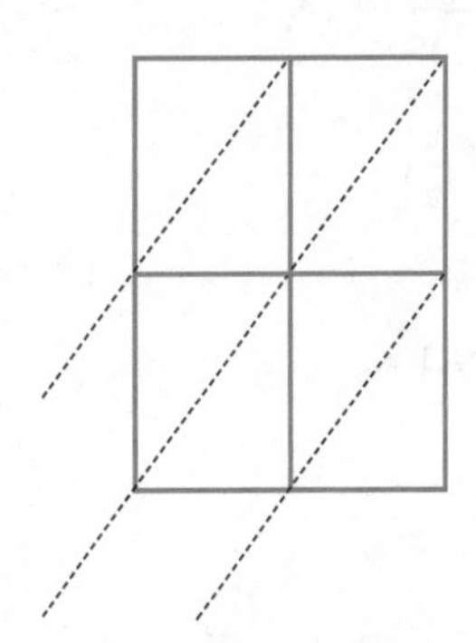

3. 384 * 256 = ____________

10s	100s	1,000s	10,000s

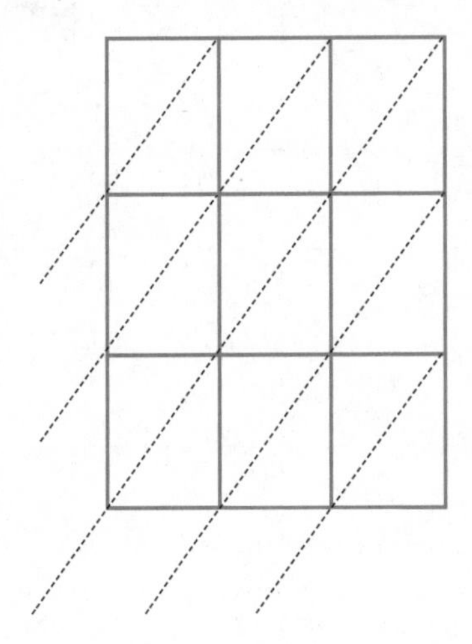

4. 6.15 * 8.3 = ____________

10s	100s	1,000s	10,000s

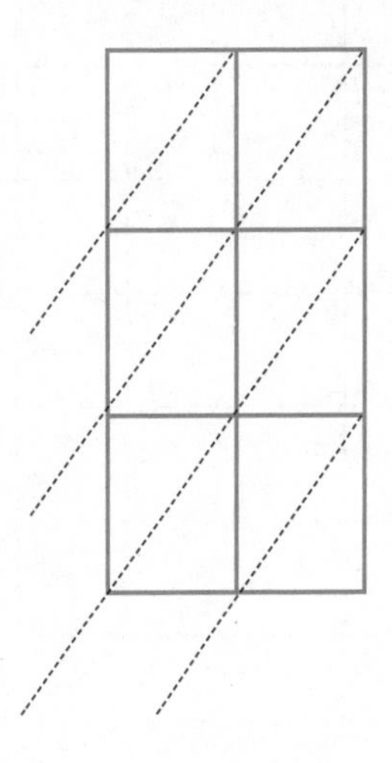

5. 1.7 * 5.6 = ____________

1s	10s	100s	1,000s

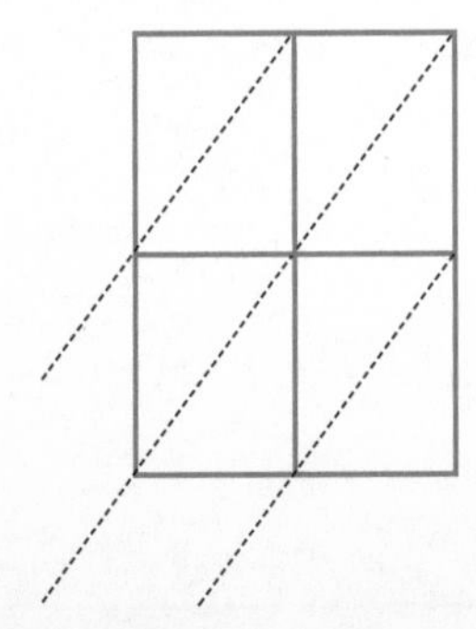

6. 82 * 4.9 = ____________

10s	100s	1,000s	10,000s

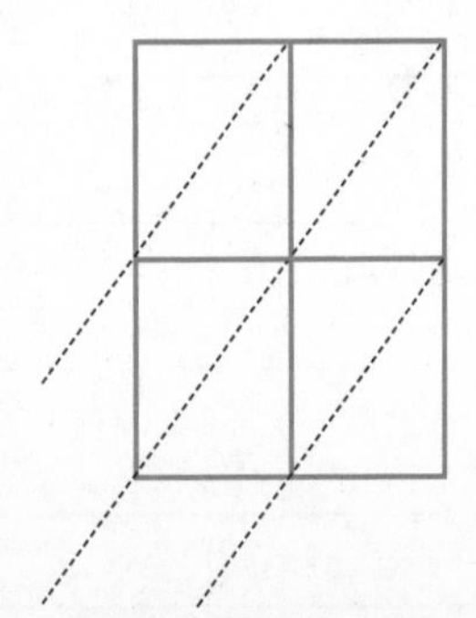

Date Time

Math Boxes 2.9

1. Cross out the shapes below that are NOT polygons.

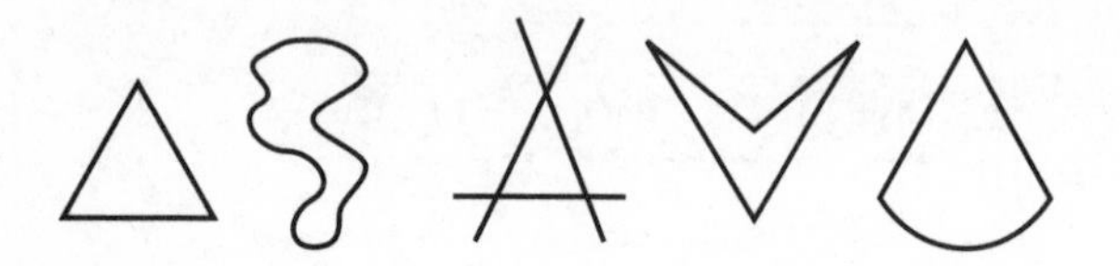

2. Find the perimeter of the polygon.

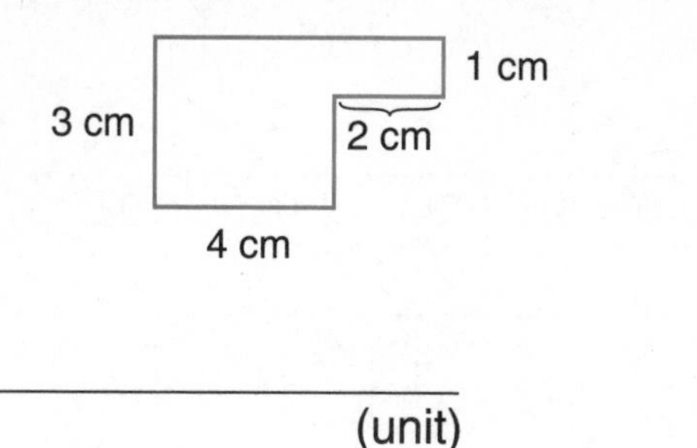

_______________ (unit)

3. Multiply. Show your work.

   a. 319 * 82 = _______________

   b. 423 * 61 = _______________

   c. _______________ = 38 * 708

   d. _______________ = 613 * 59

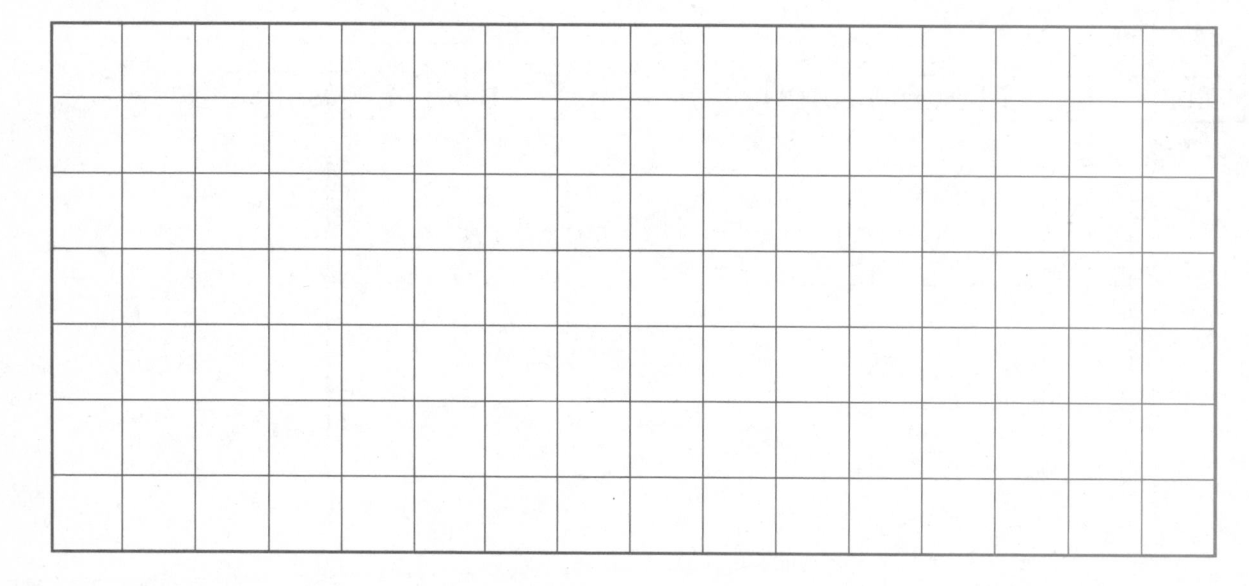

4. Give the value of the **boldface digit** in each numeral.

   a. 390.8**1** _______________

   b. **8**,092,741 _______________

   c. 4,350.**4**7 _______________

   d. 25,**3**94,008 _______________

   e. 3**2**,768.9 _______________

Date Time

Millions, Billions, and Trillions

Useful Information	
1 billion is 1,000 times 1 million. 1 million * 1 thousand = 1 billion 1,000,000 * 1,000 = 1,000,000,000	1 trillion is 1,000 times 1 billion. 1 billion * 1 thousand = 1 trillion 1,000,000,000 * 1,000 = 1,000,000,000,000
1 minute = 60 seconds 1 hour = 60 minutes 1 day = 24 hours 1 year = 365 days (366 days in a leap year)	

Make a guess: How long do you think it would take you to tap your desk 1 million times, without any interruptions? ______________

Check your guess by doing the following experiment.

1. Take a sample count.
   Record your count of taps made in 10 seconds. ____________

2. Calculate from the sample count.
   At the rate of my sample count, I expect to tap my desk:

   a. ____________ times in 1 minute.
   (*Hint:* How many 10-second intervals are there in 1 minute?)

   b. ____________ times in 1 hour.

   c. ____________ times in 1 day (24 hours).

   d. At this rate it would take me about ____________ full 24-hour days to tap my desk 1 million times.

3. Suppose that you work 24 hours per day tapping your desk. Estimate how long it would take you to tap 1 billion times and 1 trillion times.

   a. It would take me about ________________ to tap my desk 1 billion times.
   (unit)

   b. It would take me about ________________ to tap my desk 1 trillion times.
   (unit)

Date Time

Multiplication Practice

Solve the problems using your favorite multiplication method. Show your work.

1. 24 * 73 = ______________	2. 46 * 82 = ______________
3. 7.9 * 35 = ______________	4. 147 * 8 = ______________
5. 67.4 * 9.3 = ______________	6. 0.5 * 432 = ______________

Math Boxes 2.10

1. Look around the room and find an example of each of the following:

   a. parallel lines ______

   b. a rectangle ______

   c. a cylinder ______

   d. a sphere ______

   e. a trapezoid ______

2. Subtract. Show your work.

   a. 1,543 − 285 = ______

   b. \$4.48 − \$3.82 = ______

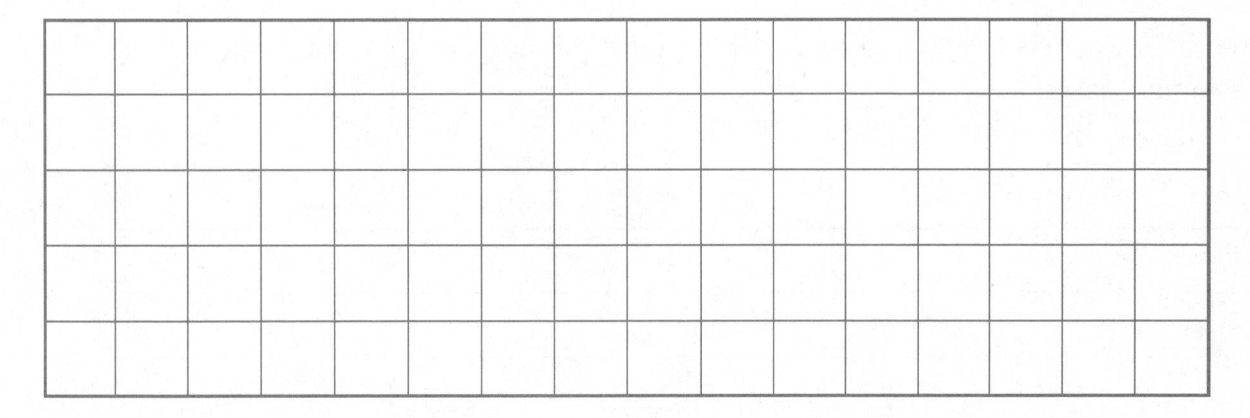

3. Use a calculator to rename each of the following in standard notation.

   a. 28^2 = ______

   b. 17^3 = ______

   c. 8^3 = ______

   d. 6^4 = ______

   e. 5^4 = ______

4. Write five names for the number 15.

   a. ______

   b. ______

   c. ______

   d. ______

   e. ______

Date Time

Time to Reflect

1. Tell which multiplication method you would use (partial-products or lattice multiplication) to find the product of 28 * 74. Explain why you favor this method.

2. What advice would you give to students working through this unit next year to help them succeed?

Math Boxes 2.11

1. I am a polygon. I have fewer sides than a quadrangle.

   a. Draw me in the space below.

   b. What shape am I? ____________

2. Measure ∠*CAT* to the nearest degree.

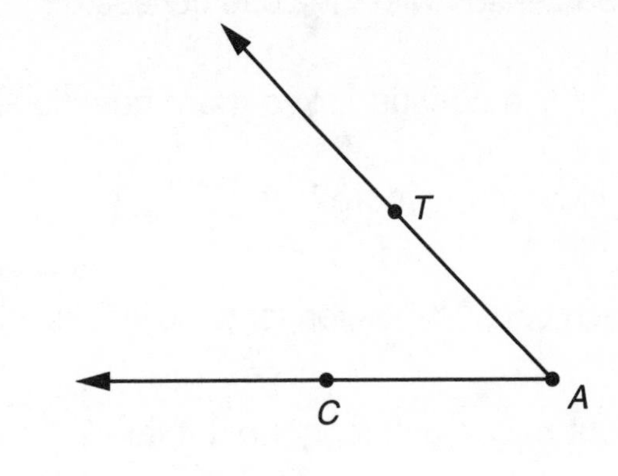

∠*CAT*: ________

3. For each shape, fill in the ovals that apply.

a.

O polygon
O parallelogram
O quadrangle
O rectangle

b. 

O polygon
O rectangle
O quadrangle
O parallelogram

c. 

O polygon
O triangle
O circle
O parallelogram

d. 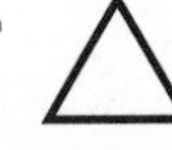

O polygon
O circle
O quadrangle
O triangle

4. Describe the attributes of a polygon. Do not use your *Student Reference Book.*

____

____

____

____

____

Date Time

U.S. Census Questions

Use the information on pages 328 and 332 of the *Student Reference Book* to compare the 1790 census with the 2000 census.

1. a. Which census asked more questions? ______________________

b. How many more? ______________________

2. Which census took longer to collect its information? ______________________

3. About how much longer did it take? ______________________

____

4. a. Which state reported the largest total population in the 1790 census?

____

b. Which state reported the smallest total population in the 1790 census?

____

5. What was the reported total population in 1790? ______________________

____

____

6. a. Were slaves counted in the 1790 Census? ______________________

b. Which state had the most slaves? ______________________

c. Which states had less than 100 slaves? ______________________

____

7. a. How many free white males were reported in Vermont in the 1790 Census?

____

b. Is this more or less than the number of free white females reported?

____

Date Time

A Mental Calculation Strategy

When you multiply a number that ends in 9, you can simplify the calculation by changing it into an easier problem. Then adjust the result.

Example 1 $2 * 99 = ?$

- Change $2 * 99$ into $2 * 100$.
- Find the answer: $2 * 100 = 200$.
- Ask, "How is the answer to $2 * 100$ different from the answer to $2 * 99$?"
  100 is 1 more than 99, and you multiplied by 2.
  So 200 is 2 more than the answer to $2 * 99$.
- Adjust the answer to $2 * 100$ to find the answer to $2 * 99$:
  $200 - 2 = 198$. So $2 * 99 = 198$.

Example 2 $3 * 149 = ?$

- Change $3 * 149$ into $3 * 150$.
- Find the answer: $3 * 150 = (3 * 100) + (3 * 50) = 450$.
- Ask, "How is the answer to $3 * 150$ different from the answer to $3 * 149$?"
  150 is 1 more than 149, and you multiplied by 3.
  So 450 is 3 more than the answer to $3 * 149$.
- Adjust: $450 - 3 = 447$. So $3 * 149 = 447$.

Use this strategy to calculate these products mentally.

1. $5 * 49 =$ ________ **2.** $5 * 99 =$ ________

3. $8 * 99 =$ ________ **4.** $4 * 199 =$ ________

5. $2 * 119 =$ ________ **6.** $3 * 98 =$ ________

Date Time

Math Boxes 3.1

1. Round 14.762 to the nearest ...

   a. tenth. ____________

   b. whole number. ____________

   c. hundredth. ____________

2. Find an object in the room that has a length of about 30 centimeters.

____

3. Solve.

a. $\begin{array}{r} 209.0 \\ -\ 73.5 \\ \hline \end{array}$

b. $\begin{array}{r} 9{,}825 \\ -\ 7{,}982 \\ \hline \end{array}$

c. $\begin{array}{r} \$30.49 \\ -\ \$8.51 \\ \hline \end{array}$

d. $\begin{array}{r} 7{,}675 \\ +\ 3{,}082 \\ \hline \end{array}$

e. $\begin{array}{r} 4.339 \\ +\ 6.671 \\ \hline \end{array}$

f. $\begin{array}{r} 5{,}946 \\ +\ 8{,}217 \\ \hline \end{array}$

4. a. Make up a set of at least twelve numbers that have the following landmarks.

Maximum: 18 Range: 13 Mode: 7 Median: 12

____

b. Make a bar graph of the data.

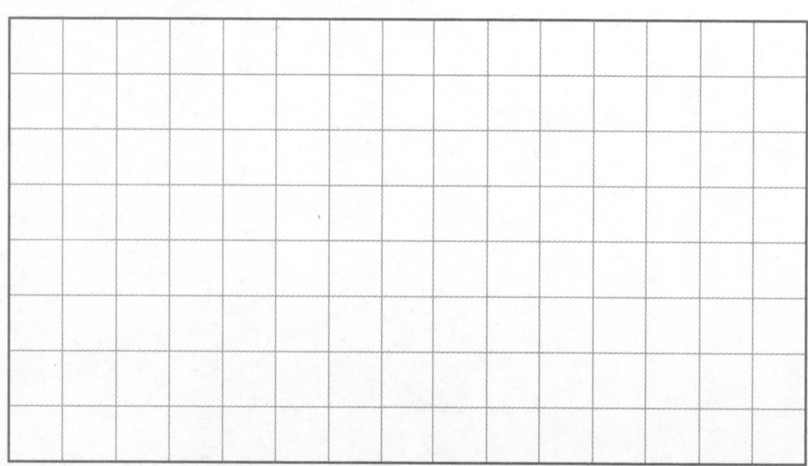

Date Time

Math Boxes 3.2

1. Complete the "What's My Rule?" table and state the rule.

Rule

in	out
20	800
3	120
40	
	2,000
	320
700	

2. Use a number line or number grid to help you subtract.

a. 24 – 30 = ________

b. 70 – 85 = ________

c. 58 – 62 = ________

d. 49 – 79 = ________

e. 90 – 104 = ________

3. Circle the best estimate for each problem.

a. 291 * 43

120   1,200   12,000

b. 68 * 32

2,100   21,000   210,000

SRB 225–228

4. Write five names for 100,000.

____

____

____

____

____

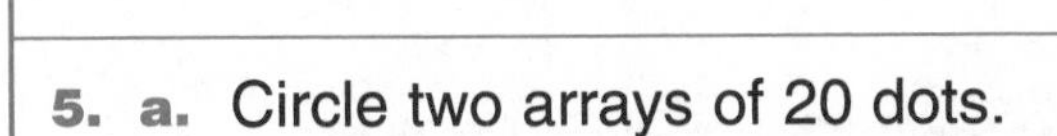

5. a. Circle two arrays of 20 dots.

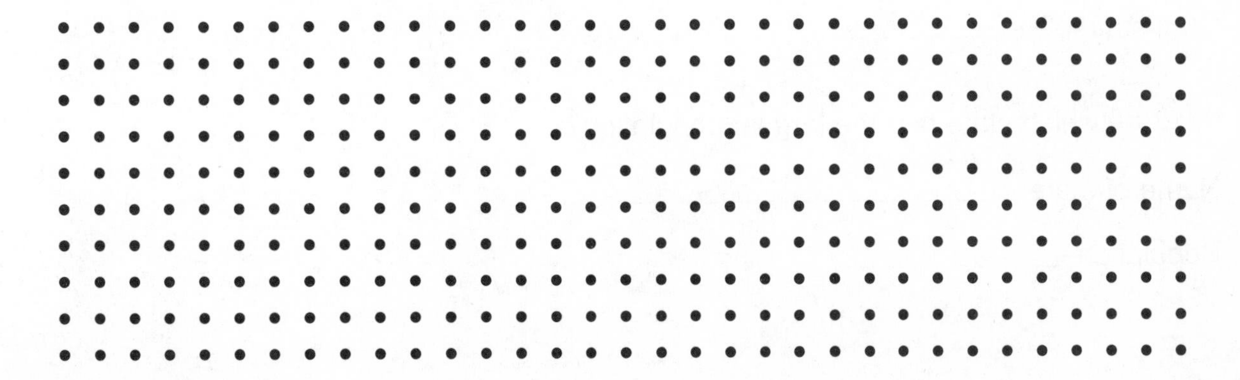

b. Write a number model for each array.

SRB 10

State Populations, 1610–1790

Use the population table on page 329 of the *Student Reference Book* to answer the following.

1. What was the population of Pennsylvania in 1780? ____________

2. What was the total population of all states in 1760? ____________

3. a. Which colony started with the smallest population?

Name of colony ____________

Year ____________

Population ____________

b. What was the population of this state in the census of 1790? ____________

4. Which colony was the first to have a population of more than 100,000?

Name of colony ____________

Year ____________

Population ____________

5. a. In what year was the total population of all states greater than 1 million for the first time? ____________

b. In what year was the total population of all states greater than 2 million for the first time? ____________

6. In 1790, which state had the largest population?

Name of state ____________

Population ____________

State Populations, 1610–1790 (cont.)

7. In 1790, which states had smaller populations than Rhode Island?

____

8. Below, fill in the total U.S. populations for 1780 and 1790. Then find how much the population increased during that 10-year period.

Population in 1790 ____________________

Population in 1780 ____________________

Increase ____________________

Challenge

9. The table gives the population of Connecticut in 1750 as 100,000. Make a mark in front of the statement that best describes the population of Connecticut in 1750.

______ It was exactly 100,000.

______ It was most likely between 99,000 and 101,000.

______ It was most likely between 95,000 and 105,000.

Explain your answer.

____

____

____

____

____

____

____

____

Date Time

Practicing Addition and Subtraction

First, estimate the answer for each problem. Then use your favorite algorithms to calculate answers for problems whose estimated sums or differences are greater than 500.

1. $\begin{array}{r} 289 \\ +\ 245 \\ \hline \end{array}$

Estimate: ________

Exact answer: ________

2. $\begin{array}{r} 1{,}013 \\ -\ 867 \\ \hline \end{array}$

Estimate: ________

Exact answer: ________

3. $\begin{array}{r} 105 \\ +\ 327 \\ \hline \end{array}$

Estimate: ________

Exact answer: ________

4. $\begin{array}{r} 941 \\ -\ 327 \\ \hline \end{array}$

Estimate: ________

Exact answer: ________

5. $\begin{array}{r} 824 \\ -\ 109 \\ \hline \end{array}$

Estimate: ________

Exact answer: ________

6. $\begin{array}{r} 214 \\ +\ 182 \\ \hline \end{array}$

Estimate: ________

Exact answer: ________

7. $\begin{array}{r} 463 \\ +\ 2{,}078 \\ \hline \end{array}$

Estimate: ________

Exact answer: ________

8. $\begin{array}{r} 1{,}532 \\ -\ 176 \\ \hline \end{array}$

Estimate: ________

Exact answer: ________

9. $\begin{array}{r} 5{,}046 \\ -\ 2{,}491 \\ \hline \end{array}$

Estimate: ________

Exact answer: ________

Date Time

Pattern-Block Angles

For each pattern block below, tell the degree measure of the angle and explain how you found the measure. Do not use a protractor.

1.

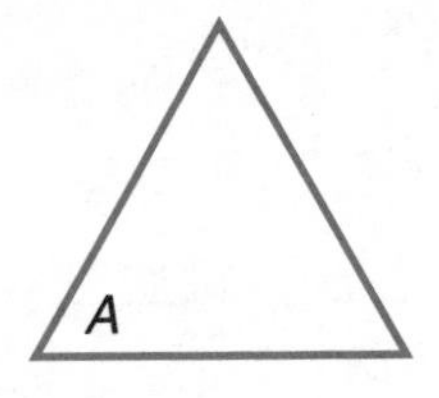

measure of $\angle A$ = ______ °

Explain. ______________________________

____

____

____

2.

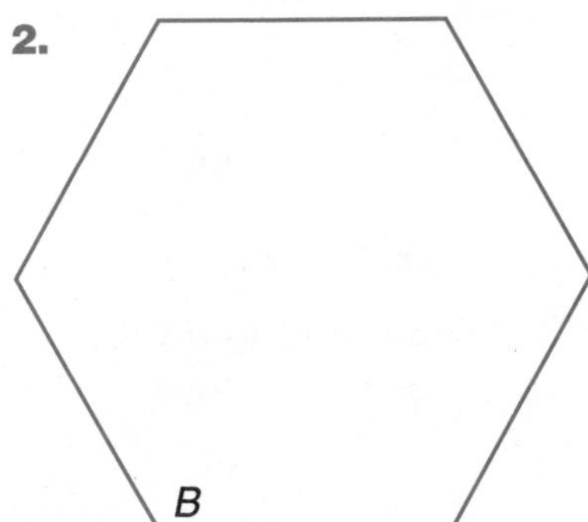

m $\angle B$ = ______ ° ("m $\angle B$" means "measure of angle *B*.")

Explain. ______________________________

____

____

____

____

3.

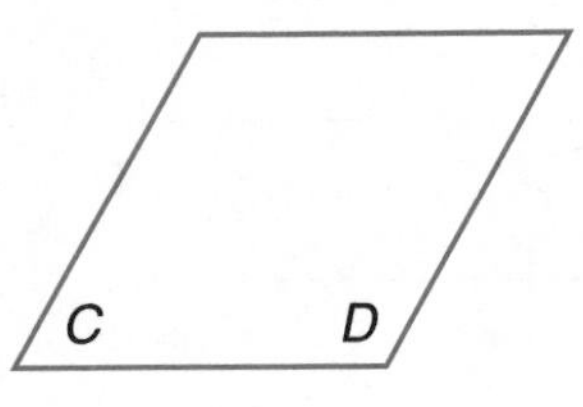

m $\angle C$ = ______ ° m $\angle D$ = ______ °

Explain. ______________________________

____

____

____

____

4.

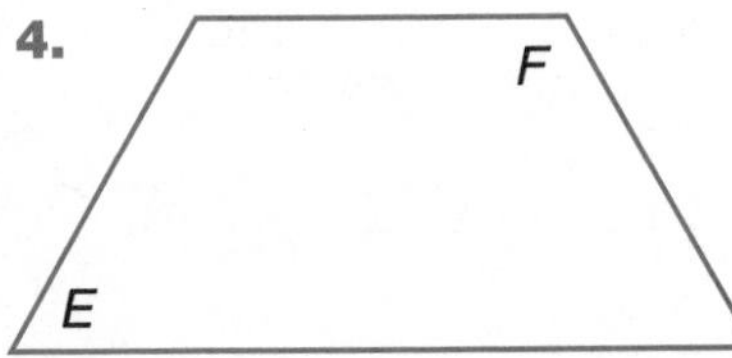

m $\angle E$ = ______ ° m $\angle F$ = ______ °

Explain. ______________________________

____

____

____

____

Date Time

Addition and Subtraction Number Stories

For each problem, fill in the blanks and solve the problem.

1. Jeanne practiced her multiplication facts for 3 weeks. The first week she practiced for 45 minutes, the second week for 37 minutes, and the third week for 32 minutes. How many minutes did she practice in all?

   a. List the numbers needed to solve the problem. ______________________

   b. Describe what you want to find. ______________________

   ____

   c. Open sentence: ______________________

   d. Solution: __________ e. Answer: ______________________ (unit)

2. The shortest book Martha read one summer was 57 pages. The longest book was 243 pages. She read a total of 36 books. How many pages longer was the longest book than the shortest book?

   a. List the numbers needed to solve the problem. ______________________

   b. Describe what you want to find. ______________________

   ____

   c. Open sentence: ______________________

   d. Solution: __________ e. Answer: ______________________ (unit)

3. Chesa collects marbles. He had 347 marbles. Then he played in two tournaments. He lost 34 marbles in the first tournament. He won 23 marbles in the second tournament. How many marbles did he have after playing in both tournaments?

   a. List the numbers needed to solve the problem. ______________________

   b. Describe what you want to find out. ______________________

   ____

   c. Open sentence: ______________________

   d. Solution: __________ e. Answer: ______________________ (unit)

Date Time

Math Boxes 3.3

1. Round 30.089 to the nearest ...

   a. tenth. ______

   b. whole number. ______

   c. hundredth. ______

2. Find an object in the room that has a length of about 18 inches.

   ______

3. Add or subtract. Show your work.

   a. 572 + 943 = ______

   b. \$15.04 + \$23.97 = ______

   c. 2,094 − 878 = ______

   d. 421.6 − 5.97 = ______

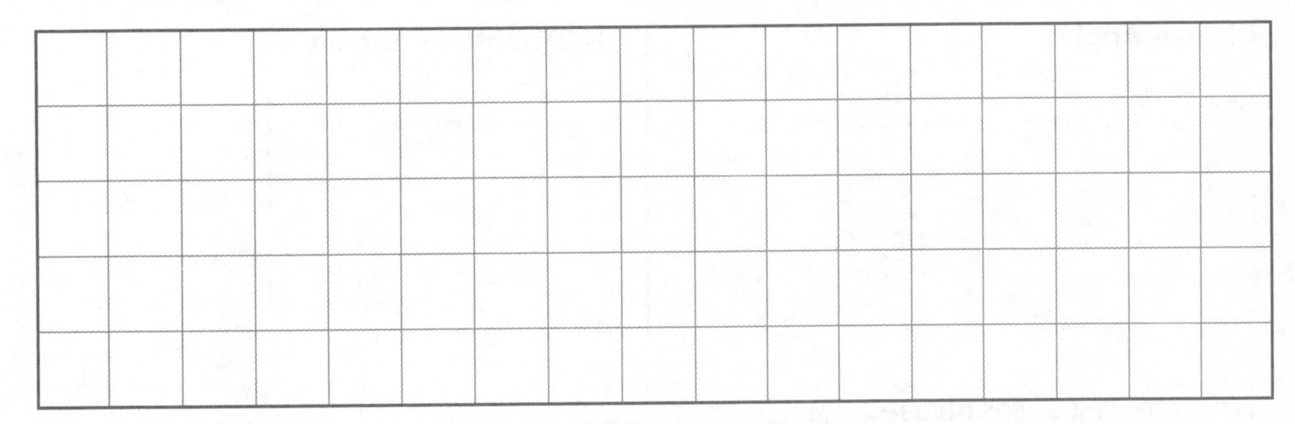

4. a. Make up a set of at least twelve numbers that have the following landmarks.

   Maximum: 8 Range: 6 Mode: 6 Median: 5

   ______

   b. Make a bar graph of the data.

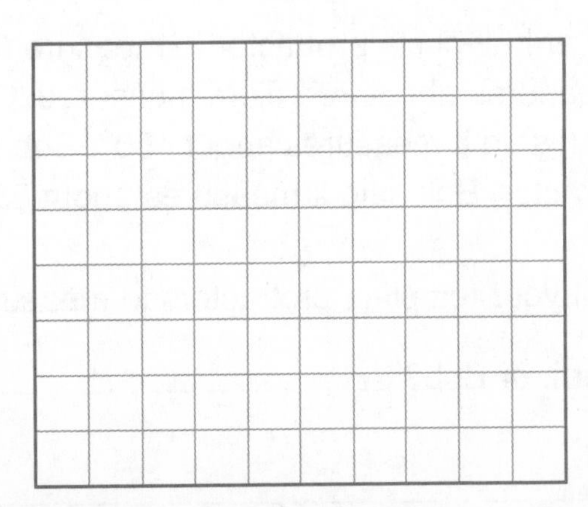

Date Time

Acute and Obtuse Angles

Math Message

1. **Acute Angles**

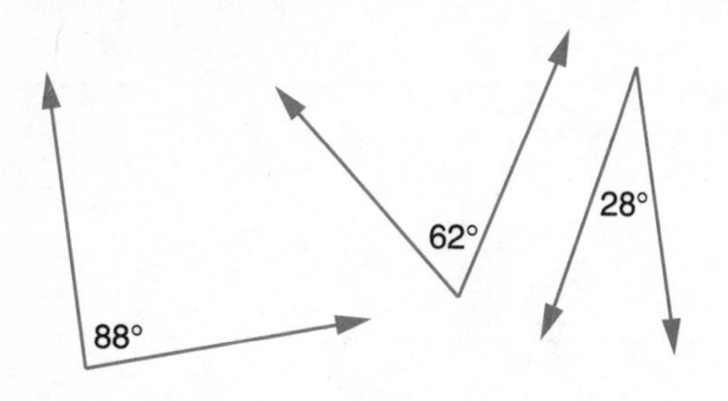

NOT Acute Angles

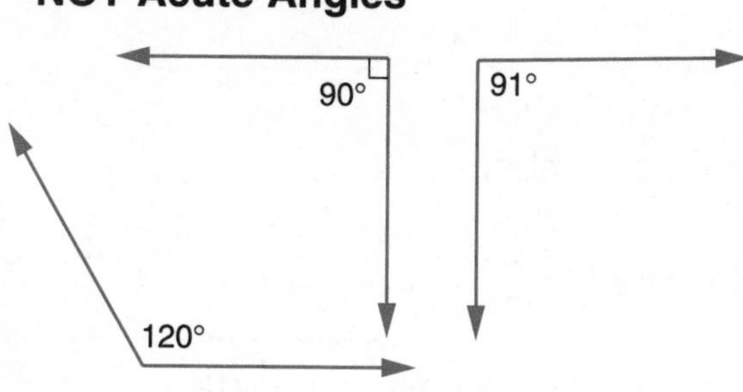

Write a definition for *acute angle.* ______________________________

____

2. **Obtuse Angles**

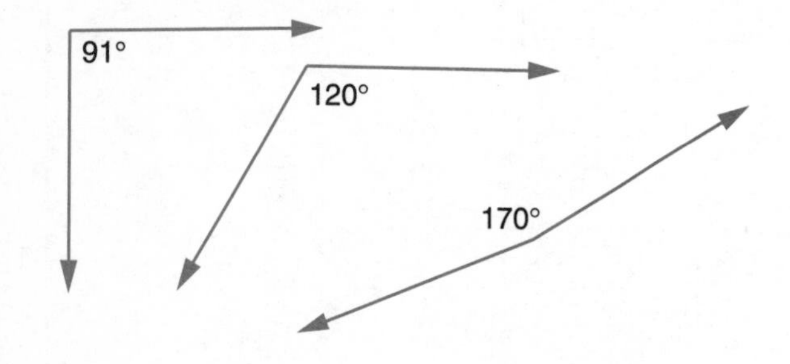

NOT Obtuse Angles

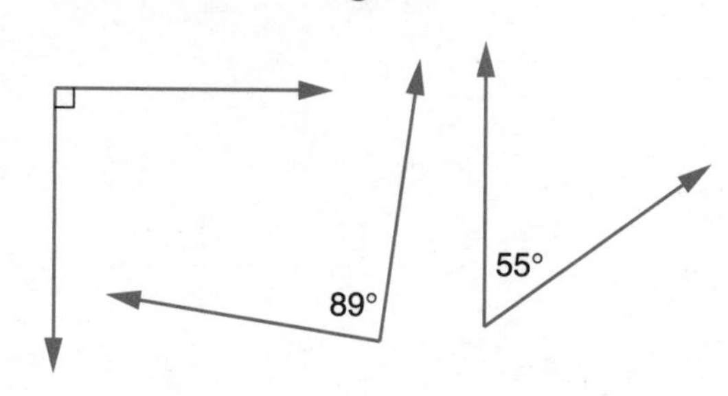

Write a definition for *obtuse angle.* ______________________________

____

Measuring and Drawing Angles with a Protractor

3. Martha used her half-circle protractor to measure the angle at the right. She said it measures about 30°. Terri measured it with her half-circle protractor. Terri said it measures about 150°. Bob measured it with his full-circle protractor. Bob said it measures about 330°.

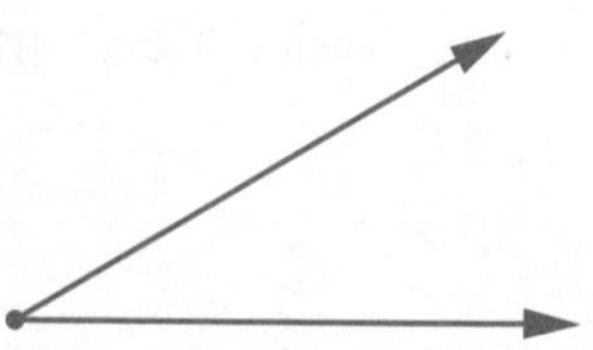

   a. Use both of your template protractors to measure the angle. Do you agree with Martha, Terri, or Bob? ______________________________

   b. Why? ______________________________

____

 Use with Lesson 3.4.

Date Time

Measuring and Drawing Angles with a Protractor (cont.)

4. Use your half-circle protractor. Measure each angle as accurately as you can.

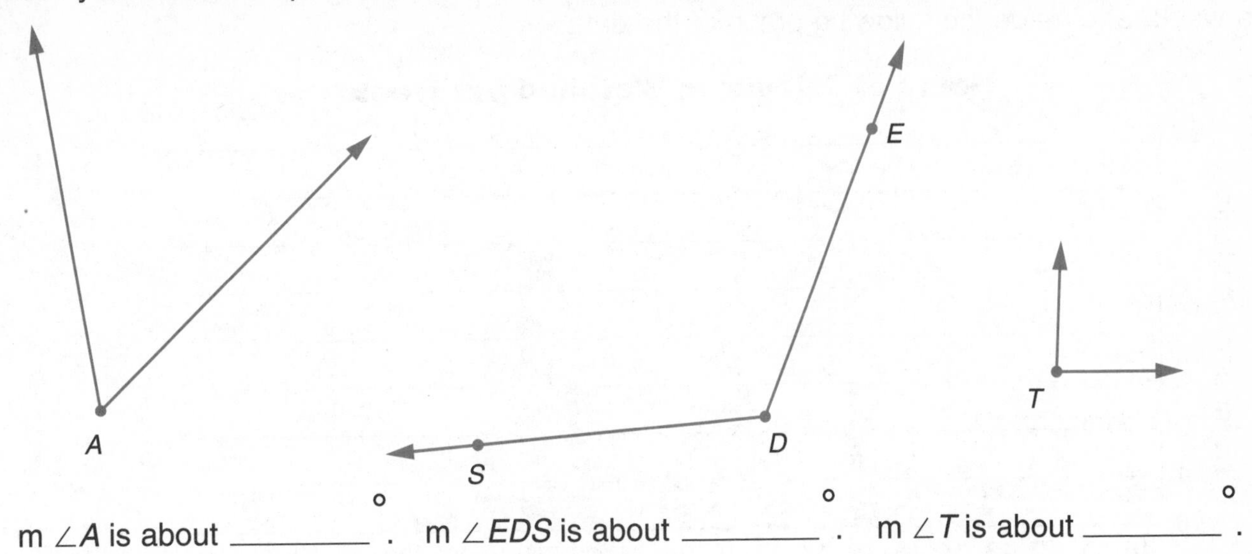

m $\angle A$ is about ________°. m $\angle EDS$ is about ________°. m $\angle T$ is about ________°.

5. Use your full-circle protractor to measure each angle.

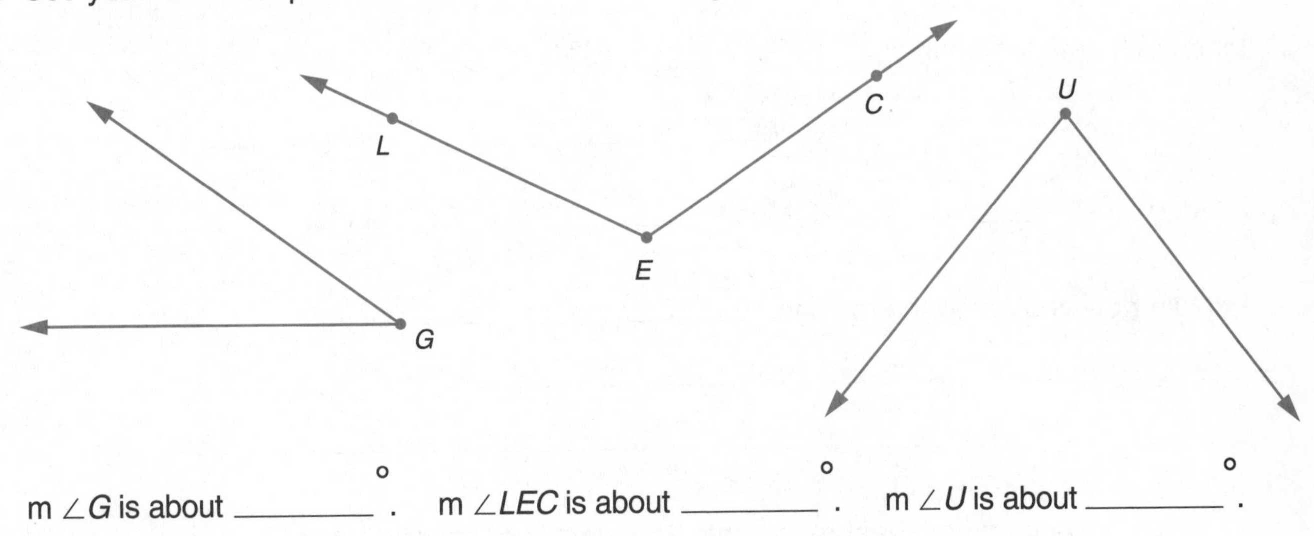

m $\angle G$ is about ________°. m $\angle LEC$ is about ________°. m $\angle U$ is about ________°.

6. Draw and label the following angles. Use your half-circle protractor.

$\angle CAT$: 62°

$\angle DOG$: 135°

Date Time

Watching Television

Adeline surveyed the students in her class to find out how much television they watch in a week. She made the following graph of the data.

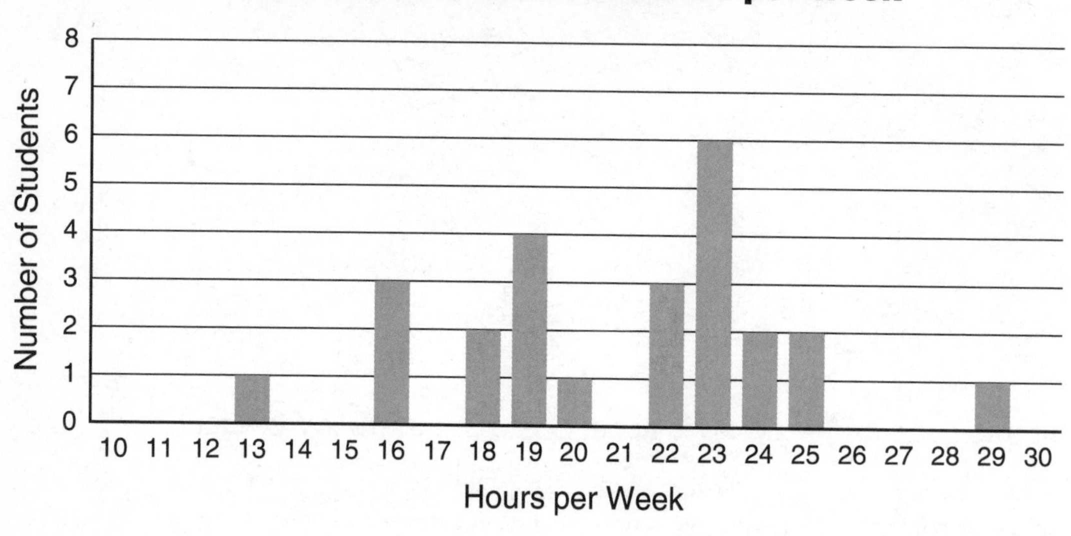

1. Find each data landmark.

a. Minimum: ________ b. Maximum: ________ c. Range: ________

d. Median: ________ e. Mean: ________ f. Mode: ________

2. Explain how you found the median. ____________________________________

____

____

3. a. Which data landmark best represents the number of hours a "typical" student watches television—the mean, median, or mode? ________________

____

b. Why? ____

____

____

Use with Lesson 3.4.

Date Time

Math Boxes 3.4

1. Complete the "What's My Rule?" table and state the rule.

Rule

in	out
40	
80	10
	9
	8
56	7

2. Use a number line or number grid to help you subtract.

a. 14 − 15 = ______

b. 25 − 32 = ______

c. 90 − 100 = ______

d. 56 − 59 = ______

e. 37 − 35 = ______

3. Circle the best estimate for each problem.

a. 38 * 47

20 200 2,000

b. 705 * 382

2,800 28,000 280,000

4. Write five names for 1,000,000.

______

______

______

______

______

5. a. Circle three different arrays of 18 dots.

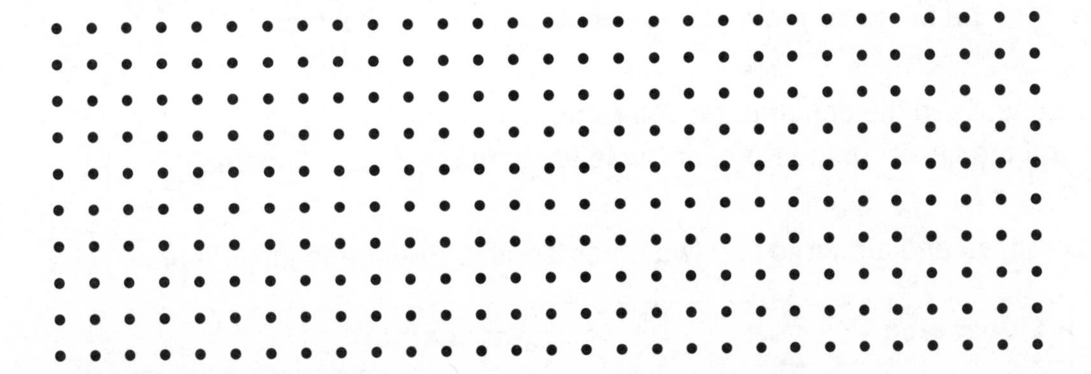

b. Write a number model for each array.

Copying Line Segments and Finding Lengths

1. Use your compass and straightedge to copy line segment AB. Do not measure the line segment with a ruler. Label the endpoints of the new line segment as points M and N. Line segment MN should be the same length as line segment AB.

2. Three line segments are shown below:

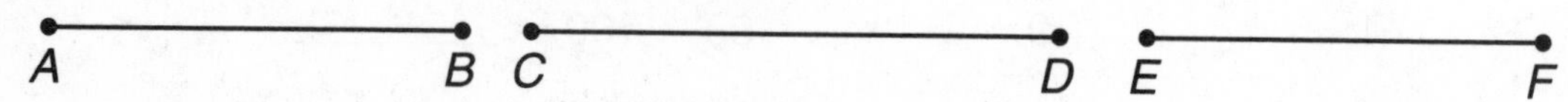

Use your compass and straightedge. Construct one line segment that is as long as the three segments joined together end to end. Label the two endpoints of the long line segment X and Y.

Use your compass to find the lengths of different parts of the Geometry Template.

Example Find the length of the longer side of the rectangle on the Geometry Template.

Step 1 Open the compass to the length of the longer side.

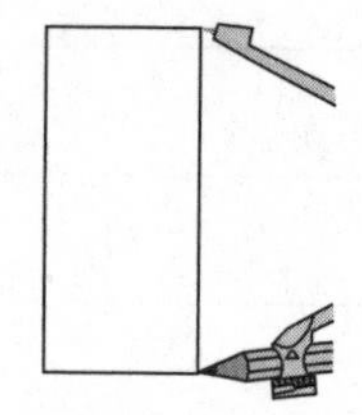

Step 2 Don't change the opening on your compass. Hold the compass against the inch ruler with the anchor at 0. Read the length. The length is about 1 inch.

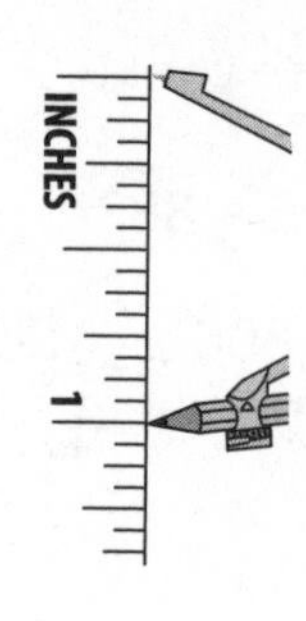

3. The length of the longer side of the trapezoid is about ________ inch(es).

4. The diameter of the full-circle protractor is about ________ inch(es).

5. The distance between the center of the full-circle protractor and the center of the Percent Circle is about ________ inch(es).

6. Use your compass and a ruler to find two other lengths. Be sure to include units.

Part Measured	Length

Date Time

Adjacent and Vertical Angles

Angles that are "next to" each other are called **adjacent angles.** Adjacent angles have the same vertex and a common side.

When two lines intersect, four angles are formed. The angles "opposite" each other are called **vertical angles** or **opposite angles.**

1. a. Angles *ABD* and *CBE* are vertical angles. Name another pair of vertical angles.

   ____

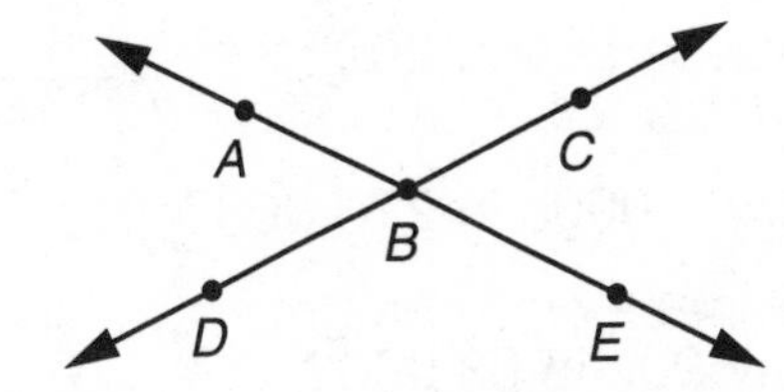

   b. Angles *ABC* and *CBE* are adjacent angles. Name two other pairs of adjacent angles.

   ____

2. The two lines at the right intersect to form four angles. One angle has been measured. Use your full-circle protractor to measure the other three angles. Record your measurements on the drawing.

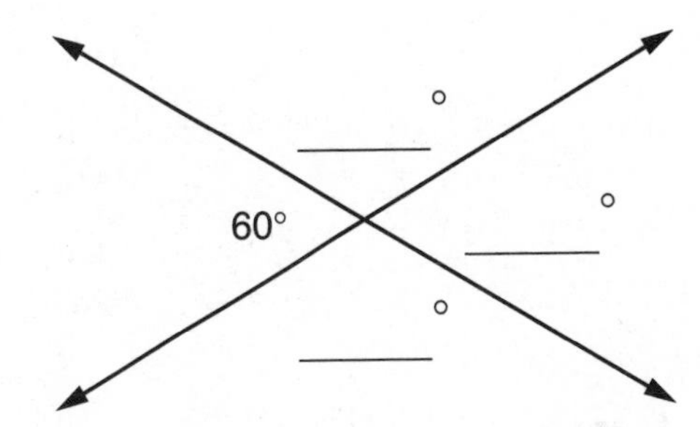

3. On a blank sheet of paper, draw two lines that intersect. Measure the four angles. Record the measures on your drawing.

4. What do you notice about the measures of pairs of vertical angles?

   ____

5. What do you notice about the measures of pairs of adjacent angles?

   ____

Challenge

6. For any pair of adjacent angles formed by two intersecting lines, the sum of the measures is always 180°. Explain why. ______________________________

   ____

Date Time

Math Boxes 3.5

1. Solve.

a. How many 80s in 7,200? ______

b. How many 600s in 54,000? ______

c. How many 5s in 450,000? ______

d. How many 3,000s in 270,000? ______

e. How many 90s in 63,000? ______

2. Write the prime factorization for 54.

______

3. Draw and label the following angle.
$\angle TOE$: 48°

4. Write a number story for the number sentence $73 * 39 = x$.
Then solve the problem.

______

______

______

______

______

______

Answer: ______

Date Time

Types of Triangles

There are small marks on the sides of some figures below. These marks show sides that are the same length. For example, in the first triangle under "Equilateral Triangles," all the sides have two marks. These sides are the same length.

For each type of triangle below, study the examples and nonexamples. Then write your own definitions. Do not use your *Student Reference Book.*

1. Equilateral Triangles

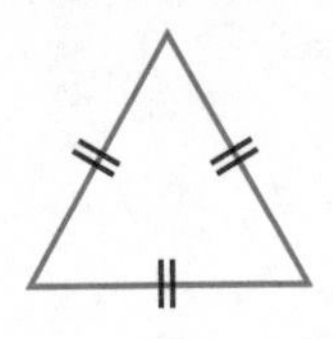

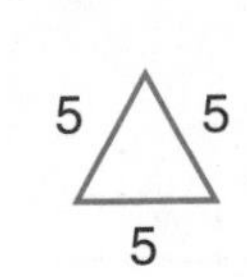

NOT Equilateral Triangles

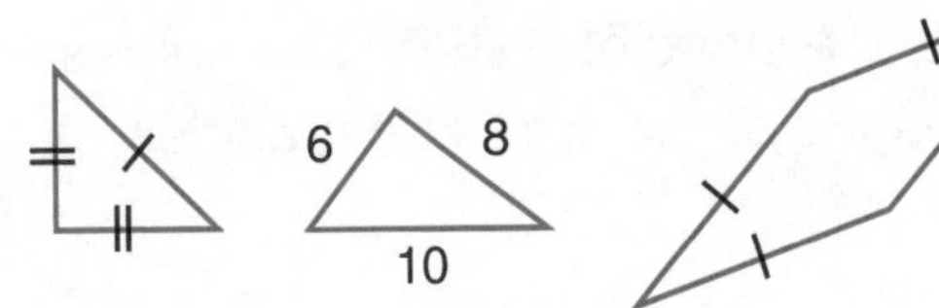

Write a definition of *equilateral triangle.* ______________________________

____

____

2. Isosceles Triangles

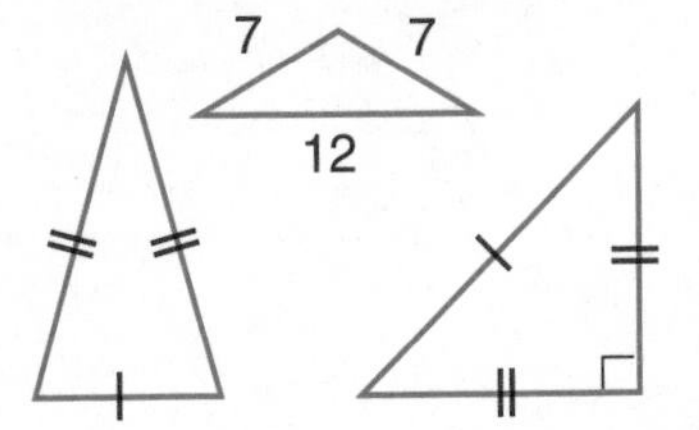

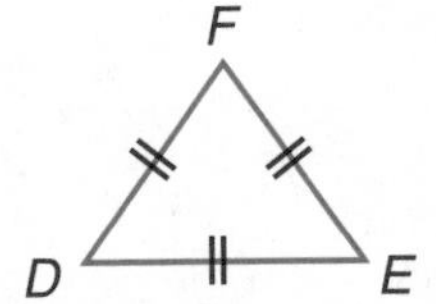

NOT Isosceles Triangles

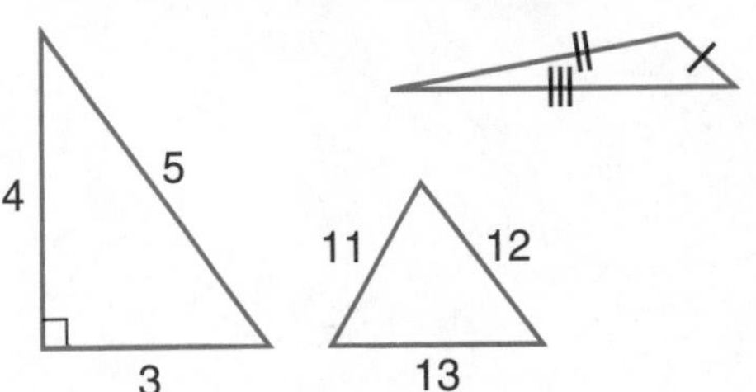

Write a definition of *isosceles triangle.* ______________________________

____

____

3. Scalene Triangles

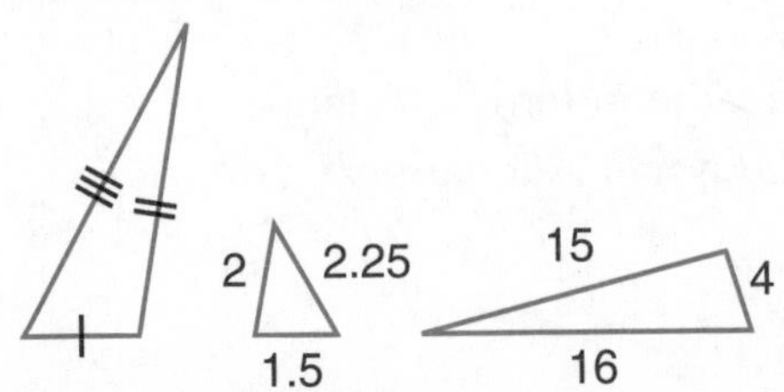

NOT Scalene Triangles

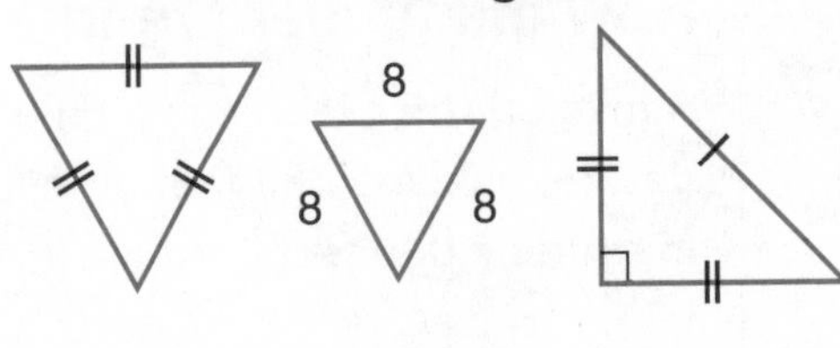

Write a definition of *scalene triangle.* ______________________________

____

Date Time

Copying a Triangle

If two triangles are identical—exactly the same size and shape—they are **congruent** to each other. Congruent triangles would match perfectly if you could move one on top of the other.

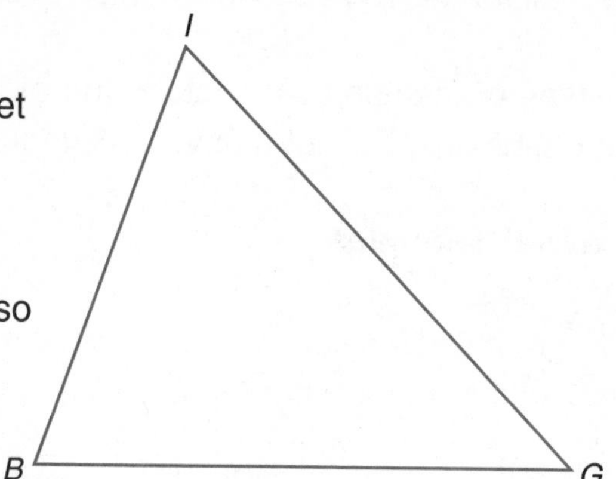

1. a. Make a copy of triangle *BIG* on a blank sheet of paper. Use any of your drawing and measuring tools, but DO NOT trace $\triangle BIG$. The sides of your copy should be the same length as the sides of $\triangle BIG$. The angles also should be the same size as the angles of $\triangle BIG$.

b. When you are satisfied with your work, cut it out and tape it in the space below. Label the vertices *P, A,* and *L*. Triangle *PAL* should be congruent to triangle *BIG*.

How many feet are in a mile?

A mile on the ocean and a mile on land are not the same in length. A land, or statute, mile is 5,280 feet. A mile on the ocean, also known as a nautical mile, measures 6,080 feet.

Source: 2201 Fascinating Facts

Date Time

Copying More Triangles

1. a. Measure the sides of triangle *HOT* in centimeters. Write the lengths next to the sides.

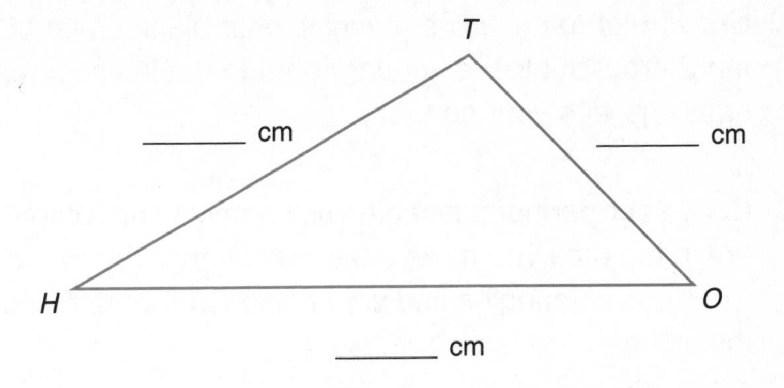

b. Make a careful copy of triangle *HOT* on a blank sheet of paper. You may use any tools EXCEPT your protractor. DO NOT trace the triangle. When you are satisfied with your work, cut it out and tape it in the space below triangle *HOT*. Label the vertices *R, E,* and *D*.

2. Make a copy of triangle *MAX* on a blank sheet of paper.

Use your compass and straightedge. DO NOT use your ruler or protractor. You may not measure the sides. When you are satisfied with your work, cut it out and tape it in the space below triangle *MAX*. Label the vertices *Y, O,* and *U*.

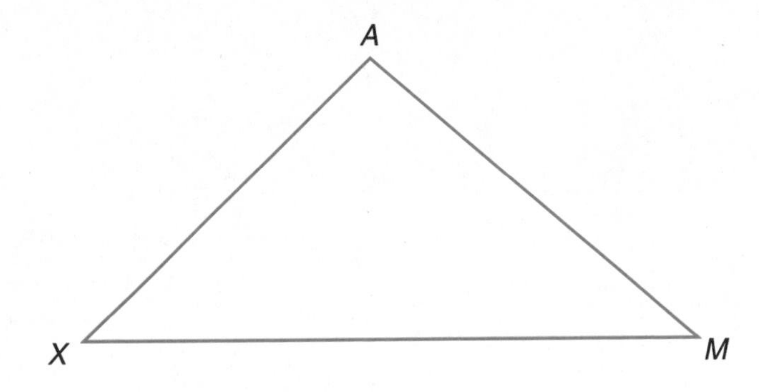

Date Time

Copying a Partner's Triangle

1. Use a ruler to draw two triangles on a blank sheet of paper. Make your triangles fairly large, but leave enough room to draw a copy of each one. Then exchange drawings with your partner.

2. Copy your partner's triangles using only your compass and straightedge. Don't erase the arcs you make—they show how you made your copies. Measure the sides of the triangles and your copies of the triangles. Write the lengths next to the sides.

3. Cut out one of the triangles your partner drew, and cut out the copy you made. Tape them in the space below.

Date Time

Math Boxes 3.6

1. Write the value of each of the following digits in the numeral 34,089,750.

   a. 4 ______________

   b. 8 ______________

   c. 5 ______________

   d. 9 ______________

   e. 3 ______________

2. I am a whole number. Use the clues to figure out what number I am.

   Clue 1 I am less than 100.

   Clue 2 The sum of my digits is 4.

   Clue 3 Half of me is an odd number.

   What number am I? ________

   Am I prime or composite?

   ____

3. True or false? Write T or F.

   a. 4,908 is divisible by 3. ________

   b. 58,462 is divisible by 2. ________

   c. 63,279 is divisible by 9. ________

   d. 27,350 is divisible by 5. ________

   e. 77,922 is divisible by 6. ________

SRB 11

4. Multiply.

   a. 30 * 900 = ______________

   b. 400 * ______________ = 40,000

   c. 800 * 6,000 = ______________

   d. 2,000 * 50 = ______________

   e. ______________ = 600 * 700

5. a. Circle the times below for which the hands on a clock form an acute angle.

   2:00 6:40 1:30 12:50

   b. Draw the hands on the clock to form an obtuse angle. (An obtuse angle measures greater than 90 degrees and less than 180 degrees.)

Date Time

Completing Partial Drawings of Polygons

Gina drew four shapes: equilateral triangle, square, rhombus, and hexagon.

She covered up most of each figure, as shown below.

Can you tell which figure is which? Write the name below each figure. Then try to draw the rest of the figure.

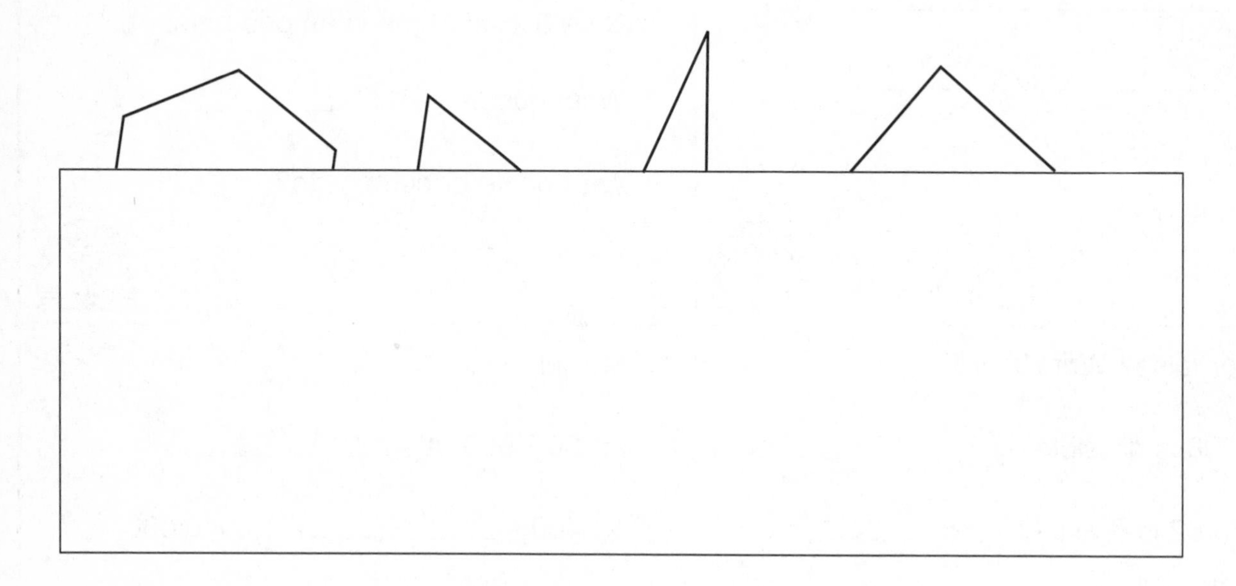

A Deep Subject

The deepest point in the world is the Mariana Trench in the Pacific Ocean. The distance from the ocean surface there to the ocean floor is about 36,000 feet—almost 7 miles. A rock the size of your head would take about an hour to fall from the surface to the ocean floor.

Source: Charlie Brown's Second Super Book of Questions and Answers

Date Time

Practicing Multiplication

First, estimate the product for each problem. Then calculate answers for problems whose estimated product is greater than 3,000.

1. 63 * 59

Estimate: ________

Exact answer: ________

2. 105 * 17

Estimate: ________

Exact answer: ________

3. 38 * 86

Estimate: ________

Exact answer: ________

4. 72 * 29

Estimate: ________

Exact answer: ________

5. 55 * 41

Estimate: ________

Exact answer: ________

6. 85 * 71

Estimate: ________

Exact answer: ________

7. 96 * 52

Estimate: ________

Exact answer: ________

8. 43 * 67

Estimate: ________

Exact answer: ________

9. 256 * 58

Estimate: ________

Exact answer: ________

Date Time

Math Boxes 3.7

1. Solve.

a. $8 * 30 =$ ____________

b. $70 *$ ____________ $= 6{,}300$

c. ____________ $* 90 = 8{,}100$

d. ____________ $= 600 * 300$

e. $800 * 5 =$ ____________

f. $400 *$ ____________ $= 20{,}000$

g. $60 * 60{,}000 =$ ____________

h. $18{,}000 =$ ____________ $* 300$

i. $45{,}000 =$ ____________ $* 90$

j. $48{,}000 =$ ____________ $* 48$

2. Write the prime factorization for 68.

____

3. Measure angle *SUM* to the nearest degree.

________ °

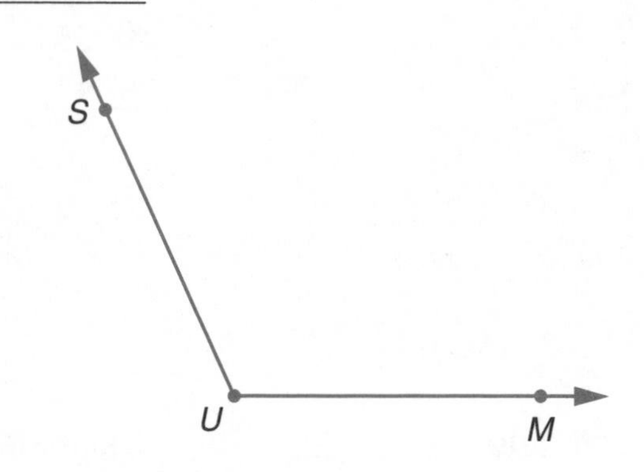

4. Write a number story for the number sentence $45 * 68 = x$.
Then solve the problem.

____

____

____

____

____

____

Answer: ____________________

Date Time

Math Boxes 3.8

1. Make the following changes to the numeral 29,078.

Change the digit
in the ones place to 4,
in the ten-thousands place to 6,
in the hundreds place to 2,
in the tens place to 9,
in the thousands place to 7.

Write the new numeral.

___ ___ , ___ ___ ___

2. I am a whole number. Use the clues to figure out what number I am.

Clue 1 I am greater than 50.

Clue 2 Half of me is less than 30.

Clue 3 My digits add up to 9.

What number am I? ________

Am I prime or composite?

____

3. True or false? Write T or F.

a. 5,894 is divisible by 6. ________

b. 6,789 is divisible by 2. ________

c. 367 is divisible by 3. ________

d. 9,024 is divisible by 4. ________

e. 8,379 is divisible by 9. ________

4. Solve.

a. 8 * 400 = ______________

b. 36,000 = ______________ * 60

c. 420,000 = 700 * ______________

d. 9,000 * ______________ = 72,000

e. 5,000 * 8,000 = ______________

5. Acute angles measure greater than 0 degrees and less than 90 degrees. Circle all the acute angles below.

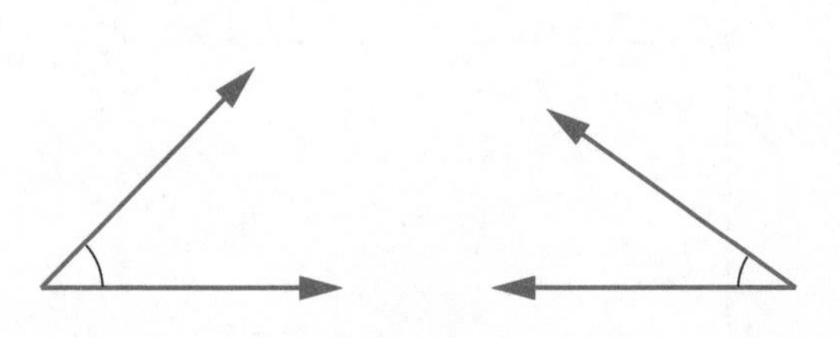

Regular Tessellations

1. A **regular polygon** is a polygon in which all sides are the same length and all angles have the same measure. Circle the regular polygons below.

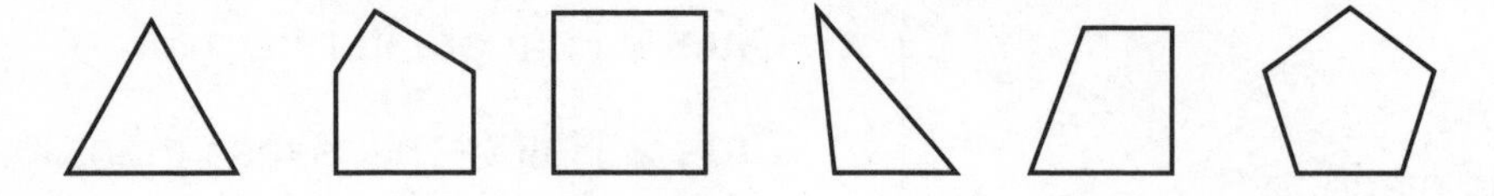

2. In the table below, write the name of each regular polygon under its picture. Then, using the polygons that you cut out from Activity Sheet 4, decide whether each polygon can be used to create a regular tessellation. Record your answers in the middle column. In the last column, use your Geometry Template to draw examples showing how the polygons tessellate or don't tessellate. Record any gaps or overlaps.

Polygon	Tessellates? (yes or no)	Draw an Example
(triangle) ________		
(square) ________		
(pentagon) ________		

Date Time

Regular Tessellations (cont.)

Polygon	Tessellates? (yes or no)	Draw an Example
________		
________		

3. Which of the polygons can be used to create regular tessellations?

____

4. Explain how you know that these are the only ones. ______________________

____

____

____

Angles in Quadrangles and Pentagons

1. Circle the kind of polygon your group is working on.

   quadrangle pentagon

2. Below, use a straightedge to carefully draw the kind of polygon your group is working on. Your polygon should look different from the ones drawn by others in your group, but it should have the same number of sides.

3. Measure the angles in your polygon. Write each measure in the angle.

4. Find the sum of the angles in your polygon. ________ °

Angles in Quadrangles and Pentagons (cont.)

5. Record your group's data below.

Group Member's Name	Sketch of Polygon	Sum of Angles

6. Find the median of the angle sums for your group. ________°

7. If you have time, draw a hexagon. Measure its angles with a protractor. Find the sum.

Sum of the angles in a hexagon = ________°

Angles in Quadrangles and Pentagons (cont.)

8. Record the class data below.

Sum of the Angles in a Quadrangle	
Group	**Group Median**

Sum of the Angles in a Pentagon	
Group	**Group Median**

9. Find the class median for each polygon. For the triangle, use the median from the Math Message.

Sums of Polygon Angles	
Polygon	**Class Median**
triangle	
quadrangle	
pentagon	
hexagon	

10. What pattern do you see in the Sums of Polygon Angles table?

____

____

____

Date Time

Angles in Heptagons

1. A heptagon is a polygon with 7 sides. Predict the sum of the angles in a heptagon. ________°

2. Draw a heptagon below. Measure its angles with a protractor. Write each measure in the angle. Find the sum.

Sum of the angles in a heptagon = ________°

3. **a.** Is your measurement close to your prediction? ______________________

b. Why might your prediction and your measurement be different?

____

____

____

____

Date Time

Angles in Any Polygon

1. Draw a line segment from vertex *A* of this octagon to each of the other vertices except *B* and *H*.

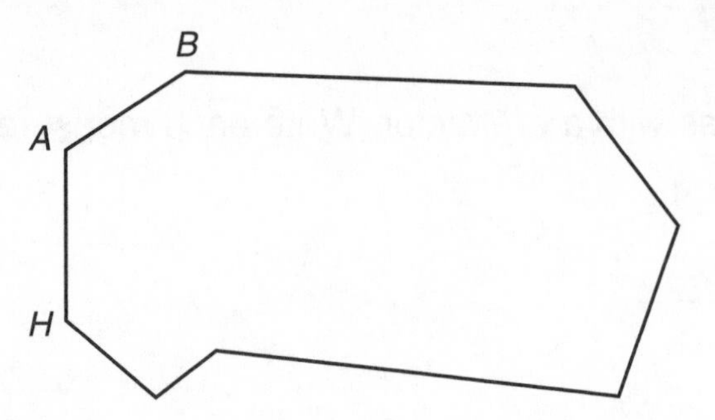

2. How many triangles did you divide the octagon into? _______________

3. What is the sum of the angles in this octagon? _______________°

4. Ignacio said the sum of his octagon's angles is 1,440°. Below is the picture he drew to show how he found his answer. Explain Ignacio's mistake.

____

____

____

____

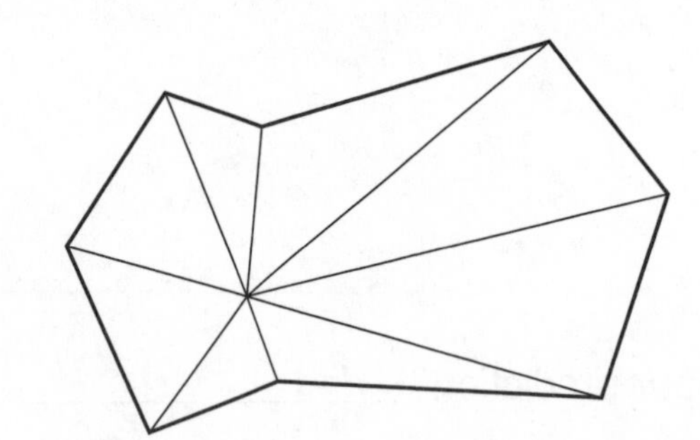

5. A 50-gon is a polygon with 50 sides. How could you find the sum of the angles in a 50-gon? ____

____

____

Sum of the angles in a 50-gon = _______________°

Date Time

Attribute Puzzles

All of these are Jimmels.	None of these are Jimmels.
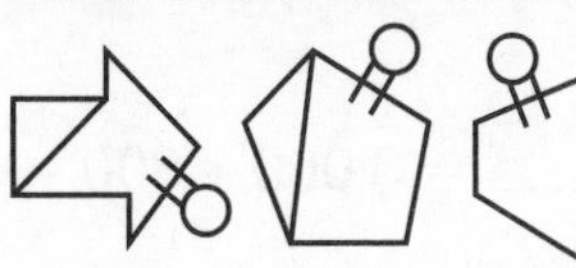 	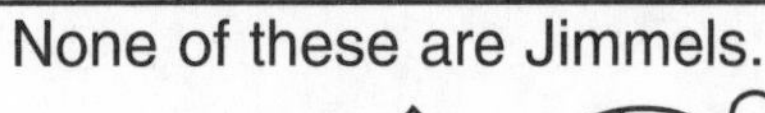 

1. List some attribute of Jimmels. ______________________________

____

____

2. Circle the Jimmels below.

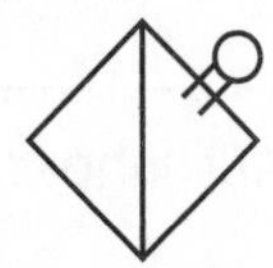 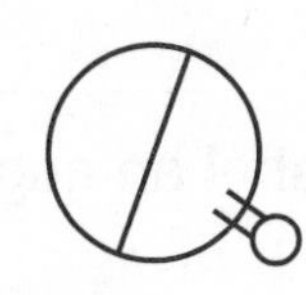   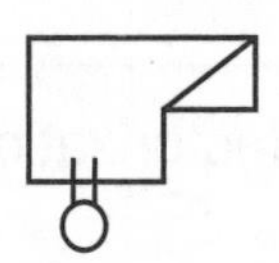

3. Draw your own Jimmel.

All of these are Dibbles.	None of these are Dibbles.
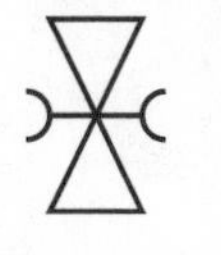 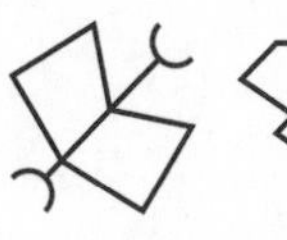 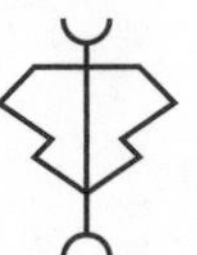 	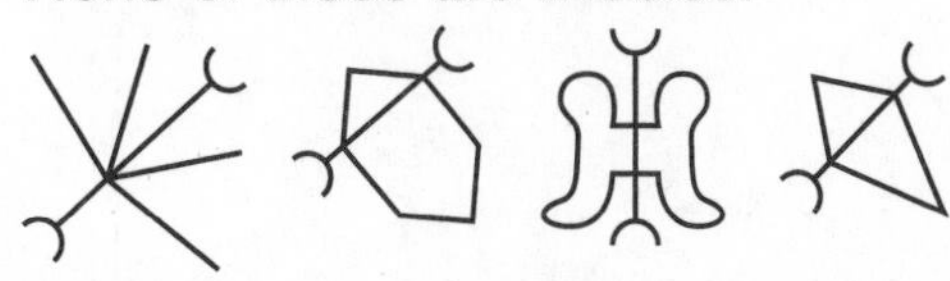

4. List some attributes of Dibbles. ______________________________

____

____

5. Circle the Dibbles below.

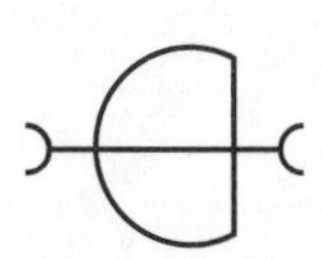 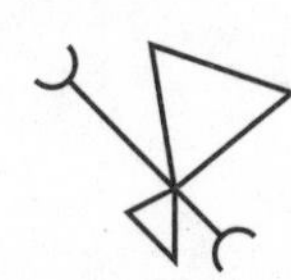  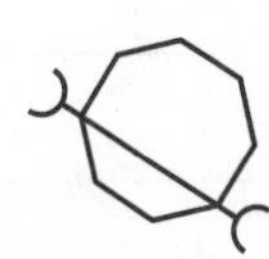 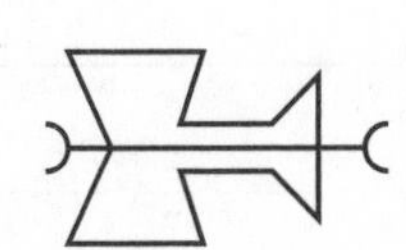

6. Draw your own Dibble.

Date Time

Math Boxes 3.9

1. Solve.

a. 3 * ______ = 270

b. ______ = 3,000 * 800

c. ______ = 500 * 400

d. ______ * 60 = 54,000

e. 60 * 50 = ______

f. ______ = 40 * 900

g. 21,000 = 700 * ______

h. 20 * 5,000 = ______

i. 800 * 600 = ______

j. 72,000 = ______ * 900

2. Write the prime factorization for 48.

______

3. Draw and label an angle *MAD,* whose measure is 105°.

4. Write a number story for the number sentence 28 * 55 = *x*. Then solve the problem.

______

______

______

______

______

______

Answer: ______

Date Time

The Geometry Template

Math Message

Answer the following questions about your Geometry Template. DO NOT count the protractors, Percent Circle, and little holes next to the rulers.

1. How many shapes are on the Geometry Template? ________

2. What fraction of these shapes are polygons? ________

3. What fraction of the shapes are quadrangles? ________

Problems for the Geometry Template

The problems on journal pages 96 and 98 are labeled Easy and Moderate. Each problem has been assigned a number of points according to its difficulty.

Complete as many of these problems as you can. Your Geometry Template and a sharp pencil are the only tools you may use. Record and label your answers on the page opposite the problems.

Some of the problems may seem confusing at first. Before asking your teacher for help, try the following:

- Look at the examples on the journal page. Do they help you understand what the problem is asking you to do?
- If you are not sure what a word means, look it up in the Glossary in your *Student Reference Book.* You might also look for help in the geometry section of the *Student Reference Book.*
- Find a classmate who is working on the same problem. Can the two of you work together to find a solution?
- Find a classmate who has completed the problem. Can she or he give you hints about how to solve it?

When the time for this activity has ended, you may want to total the number of points that you have scored. If you didn't have time to complete all these pages, you can continue working on them when you have free time.

Good luck and have fun!

Date Time

Problems for the Geometry Template (cont.)

Record your solutions on journal page 97. Include the problem numbers.

Easy **Examples**

1. Using only shapes on your Geometry Template, draw an interesting picture. (2 points)

2. Trace all of the polygons on the Geometry Template that have at least one pair of **parallel sides.** (1 point each)

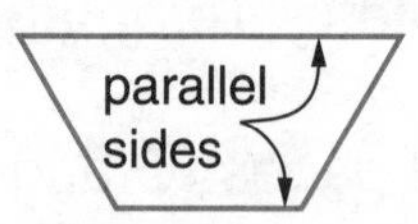

3. Trace all of the polygons on the Geometry Template that have no pairs of parallel sides. (1 point each)

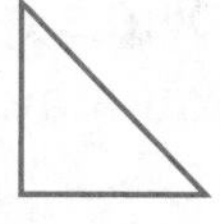

4. Trace three polygons that have *at least* one **right angle** each, three polygons that have *at least* one **acute angle** each, and three polygons that have *at least* one **obtuse angle** each. ($\frac{1}{2}$ point each)

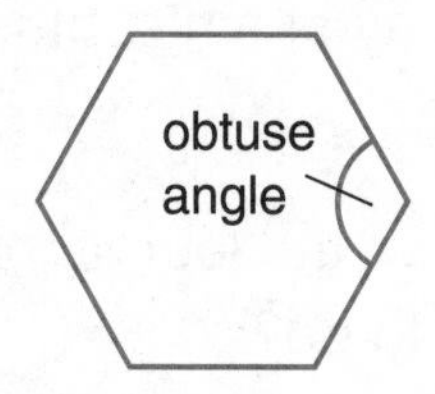

5. Assume that the side of the largest square on the template has a length of 1 **unit.** Draw three different polygons, each with a **perimeter** of 8 units. (2 points each)

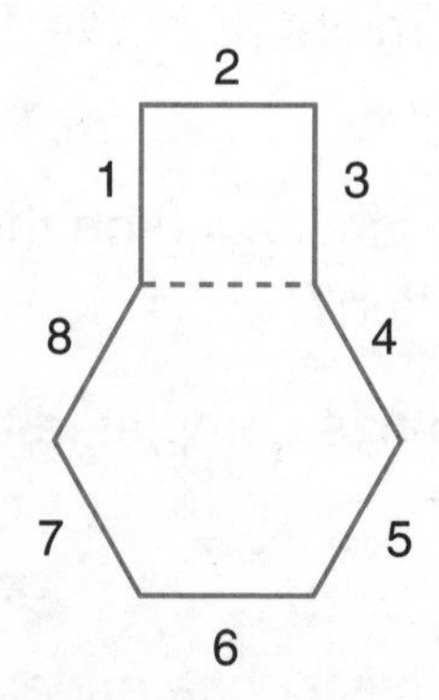

Date Time

Problems for the Geometry Template (cont.)

Solutions

Problems for the Geometry Template (cont.)

Record your solutions on journal page 99. Include the problem numbers.

Moderate **Examples**

6. Use your template to **copy** this design. (3 points)

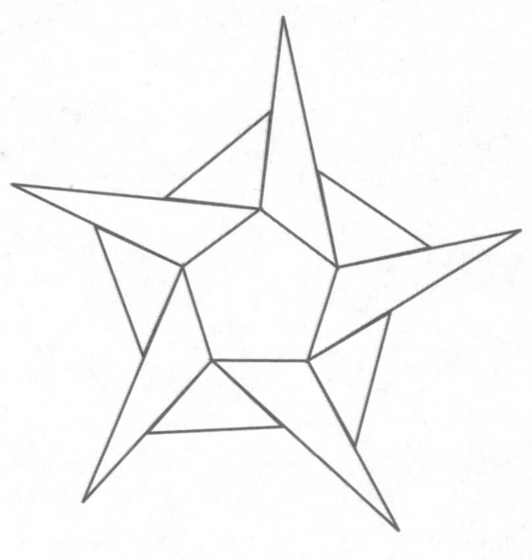

7. Without using a ruler, enlarge the rectangle. First, draw a rectangle twice the size of the rectangle on the Geometry Template. Then, draw a rectangle 3 times the size of the rectangle on the Geometry Template. (3 points each)

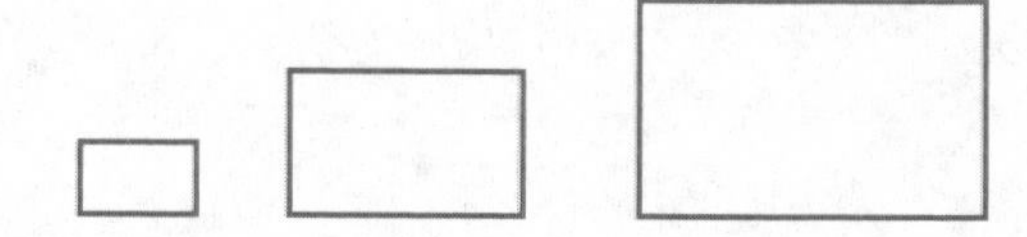

8. Compare the **perimeters** of the rectangle and the pentagon on the Geometry Template. Which polygon has the greater perimeter? You may not use the rulers on the template to help you. Describe how you were able to use other parts of your Geometry Template to solve this problem. (6 points)

9. Use the triangles on the Geometry Template to draw four different **parallelograms.** (2 points each)

10. Using any two polygons from the Geometry Template, draw five different **pentagons.** (2 points each)

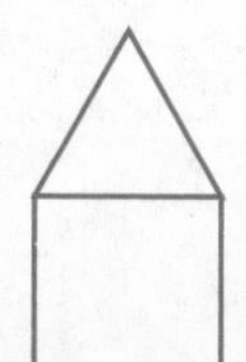

Date Time

Problems for the Geometry Template (cont.)

Solutions

Date Time

Math Boxes 3.10

1. Make the following changes to the numeral 34,709.

Change the digit
in the ones place to 6,
in the tens place to 5,
in the thousands place to 0,
in the ten-thousands place to 9,
in the hundreds place to 3.

Write the new numeral.

___ ___ , ___ ___ ___

2. I am a whole number. Use the clues to figure out what number I am.

Clue 1 Half of me is greater than 20.

Clue 2 One of my digits is double the other.

Clue 3 One of my digits is a perfect square.

What number am I? ________

Am I prime or composite?

____

3. True or false? Write T or F.

a. 1,704 is divisible by 4. ________

b. 7,152 is divisible by 6. ________

c. 8,264 is divisible by 3. ________

d. 4,005 is divisible by 2. ________

e. 2,793 is divisible by 9. ________

4. Solve.

a. 8 * 700 = ____________

b. 36,000 = ____________ * 40

c. 320,000 = 800 * ____________

d. 2,000 * ____________ = 24,000

e. 5,000 * 4,000 = ____________

5. a. Use a straightedge to draw an angle that is greater than 90°.

b. Use a straightedge to draw an angle that is less than 90°.

Time to Reflect

1. Look back through your journal. Then describe what you liked most in this unit.

____

____

____

____

____

____

2. This unit was about geometry. Based on the lessons you did in this unit, how would you describe geometry to someone?

____

____

____

____

____

____

Date Time

Math Boxes 3.11

1. Round 50.92 to the nearest …

   a. tenth. _______________

   b. whole number. _______________

   c. ten. _______________

2. Complete the "What's My Rule?" table and state the rule.

Rule

in	out
240	8
600	20
	12
	50
2,100	
1,200	

3. Solve.

   a. How many 90s in 450? __________

   b. How many 700s in 2,100? __________

   c. How many 60s in 5,400? __________

   d. How many 5s in 35,000? __________

   e. How many 80s in 5,600? __________

4. Make the following changes to the numeral 6,205.12.

   Change the digit
   in the ones place to 7,
   in the hundreds place to 5,
   in the tenths place to 6,
   in the tens place to 8,
   in the thousands place to 4.

   Write the new numeral.

   ___ , ___ ___ ___ . ___ ___

5. Circle the best estimate for each problem.

   a. 522 * 397

   2,000   20,000   200,000

   b. 1,483 * 23

   3,000   30,000   300,000

Date Time

A Mental Division Strategy

If you want to divide 56 by 7 in your head, think: *How many 7s are there in 56?* or *7 times what number equals 56?*

Since $7 * 8 = 56$, you know that there are 8 [7s] in 56. So, 56 divided by 7 equals 8.

Fact knowledge can also help you find how many times a 1-digit number will divide any number. Just break the larger number into two or more "friendly" numbers—numbers that are easy to divide by the 1-digit number.

Example 1 96 divided by 3

Break 96 into smaller, "friendly" numbers, such as the following:

- 90 and 6. Ask yourself: *How many 3s in 90?* (30) *How many 3s in 6?* (2) Total = 30 + 2 = 32.
- 60 and 36. Ask yourself: *How many 3s in 60?* (20) *How many 3s in 36?* (12) Total = 20 + 12 = 32.

So, 96 divided by 3 equals 32. Check the result: $3 * 32 = 96$.

Example 2 How many 4s in 71?

Break 71 into smaller, "friendly" numbers, such as the following:

- 40 and 31. Ask yourself: *How many 4s in 40?* (10) *How many 4s in 31?* (7 and 3 left over) (Think: *What multiplication fact for 4 has a product near 31?* $4 * 7 = 28$.) Total = 17 and 3 left over.
- 20, 20, 20, and 11. Ask yourself: *How many 4s in 20?* (5) *How many 4s in three 20s?* (15) *How many 4s in 11?* (2 and 3 left over) Total = 17 and 3 left over.

So, 71 divided by 4 equals 17 with 3 left over.

Use this method to mentally find or estimate the following. Remember to break the number being divided into two or more friendly parts.

1. 42 divided by 3 equals ______.

   ____
   (friendly parts for 42)

2. 57 divided by 3 equals ______.

   ____
   (friendly parts for 57)

3. 96 divided by 8 equals ______.

   ____
   (friendly parts for 96)

4. 99 divided by 7 equals ______.

   ____
   (friendly parts for 99)

Date Time

Place-Value Puzzles

1. The digit in the thousands place is 6.

   The digit in the ones place is the sum of the digits in a dozen.

   The digit in the millions place is $\frac{1}{10}$ of 70.

   The digit in the hundred-thousands place is $\frac{1}{2}$ of the digit in the thousands place.

   The digit in the hundreds place is the sum of the digit in the thousands place and the digit in the ones place.

   The rest of the digits are all 5s. ___ ___, ___ ___ ___, ___ ___ ___

2. The digit in the tens place is 2.

   The digit in the ones place is double the digit in the tens place.

   The digit in the hundreds place is three times the digit in the tens place.

   The digit in the hundred-thousands place is an odd number less than 3.

   The digit in the millions place is $\frac{1}{3}$ of 15.

   The rest of the digits are all 9s. ___ ___, ___ ___ ___, ___ ___ ___

3. The digit in the ten-thousands place is the sum of the digits in 150.

   The digit in the millions place is a prime number greater than 5.

   The digit in the hundreds place is $\frac{1}{2}$ of the digit in the thousands place.

   The digit in the tenths place is 1 less than the digit in the millions place.

   The digit in the thousands place is $\frac{2}{5}$ of 20.

   The rest of the digits are all 3s. ___, ___ ___ ___, ___ ___ ___ . ___ ___

Challenge

4. The digit in the thousands place is the smallest square number greater than 1.

   The digit in the tens place is the same as the digit in the place 1,000 times greater.

   The digit in the ten-thousands place is $\frac{1}{2}$ of the digit in the ten-millions place.

   The digit in the ten-millions place is two more than the digit in the thousands place.

   The digit in the hundreds place is 1 greater than double the digit in the ten-thousands place.

   The rest of the digits are all 2s. ___ ___, ___ ___ ___, ___ ___ ___

Date Time

Math Boxes 4.1

1. Measure each line segment to the nearest quarter-inch.

   a. ____________

   ________ in.

   b. ______

   ________ in.

2. I have 4 sides and 2 acute angles. All of my opposite sides are parallel. What shape am I?

   ____

3. Estimate the answer to each multiplication problem.

   a. 303 ∗ 78 = ____________

   b. 49 ∗ 59 = ____________

   c. 23 ∗ 99 = ____________

   d. 607 ∗ 12 = ____________

   e. 91 ∗ 91 = ____________

4. Round 7,403.93 to the nearest ...

   a. hundred. ____________

   b. tenth. ____________

   c. whole number. ____________

5. Roger had saved $10.05 from his allowance. Then he bought a paint-by-numbers kit for $7.39. How much does he have left?

   ____

Date Time

The Partial-Quotients Division Algorithm

These notations for division are equivalent:

$246 / 12 = ?$ $\frac{246}{12} = ?$ $12\overline{)246}$ (quotient ?) $246 \div 12 = ?$

Here is an example of division using the partial-quotients algorithm:

$8\overline{)185}$		How many 8s are in 185? At least 10.
− 80	10	The first partial quotient. 10 * 8 = 80
105		Subtract. At least 10 [8s] are left.
− 80	10	The second partial quotient. 10 * 8 = 80
25		Subtract. At least 3 [8s] are left.
− 24	3	The third partial quotient. 3 * 8 = 24
1	23	Subtract. Add the partial quotients: 10 + 10 + 3 = 23
↑	↑	
Remainder	**Quotient**	**Answer: 23 R1**

Divide.

1. $8\overline{)264}$ ______________
2. $749 \div 7$ ______________
3. $2{,}628 / 36$ ______________
4. Raoul has 237 string bean seeds. He plants them in rows with 8 seeds in each row. How many complete rows can he plant? __________ rows

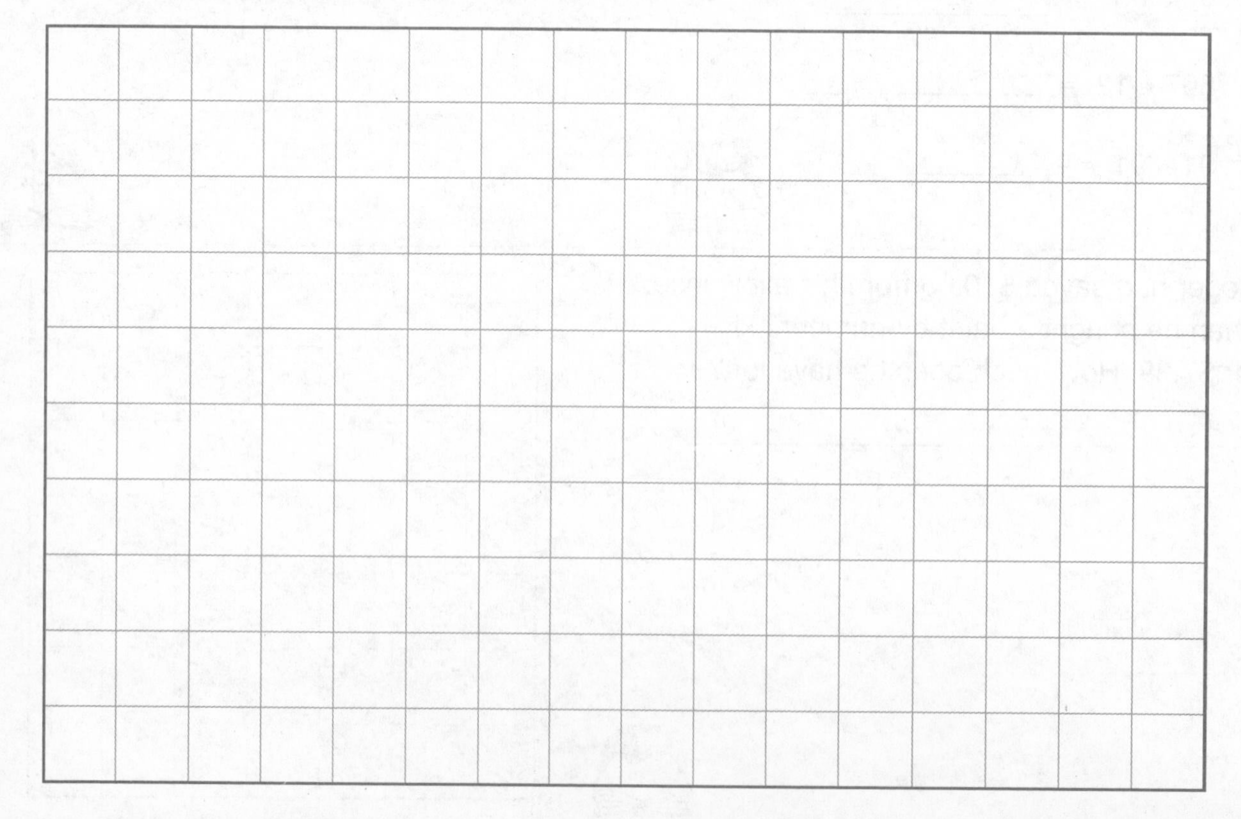

Date Time

The Partial-Quotients Division Algorithm (cont.)

Divide.

5. 823 / 3 ____________ 6. $43\overline{)2{,}815}$ ____________ 7. $\frac{4{,}290}{64}$ ____________

8. Regina put 1,610 math books into boxes. Each box held 24 books. How many boxes did she fill? ____________ boxes

9. Make up a number story that can be solved with division. Solve it using a division algorithm.

____

____

____

Solution: ____

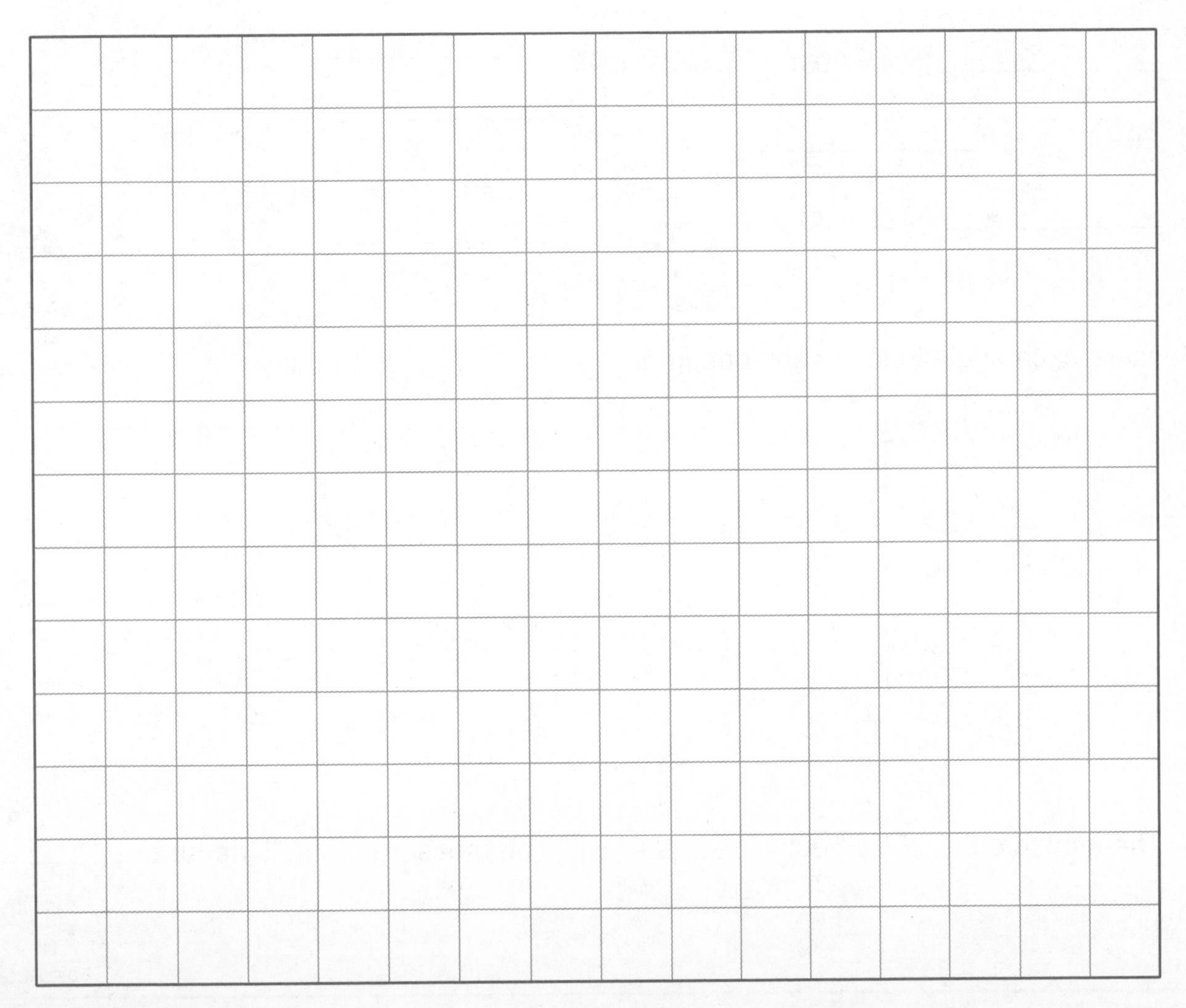

Date Time

Math Boxes 4.2

1. Measure the length and width of each of the following objects to the nearest half inch.

a. piece of paper length ________ in. width ________ in.

b. dictionary length ________ in. width ________ in.

c. palm of your hand length ________ in. width ________ in.

d. ________________ (your choice) length ________ in. width ________ in.

2. Do the following multiplication problems mentally.

a. 89 * 5 = ____________

b. 199 * 12 = ____________

c. ____________ = 4 * 399

d. 29 * 15 = ____________

e. ____________ = 59 * 30

3. Write the following numbers in order from least to greatest.

2.05 2.70 2.57 2.07 2.5

____

____

____

____

____

4. Measure each angle to the nearest degree.

a.

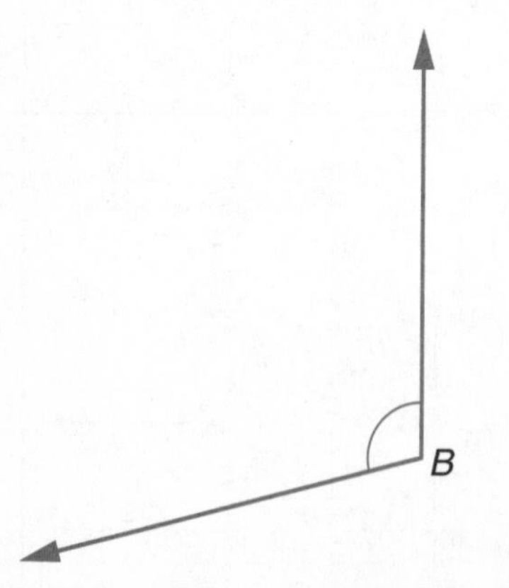

The measure of $\angle B$ is about ________ °.

b.

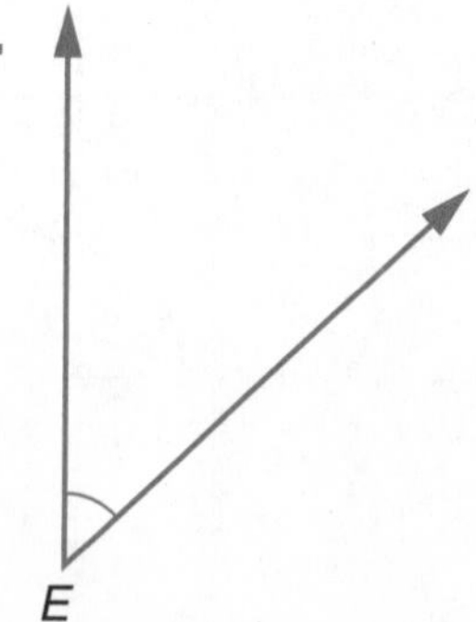

The measure of $\angle E$ is about ________ °.

SRB 188 189

Date Time

Distances between U.S. Cities

1. Use the map of the United States on pages 344 and 345 of your *Student Reference Book* to estimate the distances between the following cities. Measure each map distance in inches. Complete the table. (Scale: 1 inch represents 200 miles)

Cities	Map Distance (inches)	Real Distance (miles)
Chicago, IL, to Pittsburgh, PA	2 inches	400 miles
Little Rock, AR, to Jackson, MS		
San Francisco, CA, to Salt Lake City, UT		
Indianapolis, IN, to Raleigh, NC		
Chicago, IL, to Boston, MA		
San Antonio, TX, to Buffalo, NY		
Salt Lake City, UT, to Pierre, SD		

2. Explain how you found the real distance from Salt Lake City, UT, to Pierre, SD.

3. Explain who might use a map scale and why.

Date Time

Measuring Paths That Are Not Straight

Use a ruler, string, compass, paper and pencil, or any other tool.

1. The map below shows the border between Mexico and the United States. Estimate the length of the border. ______ miles

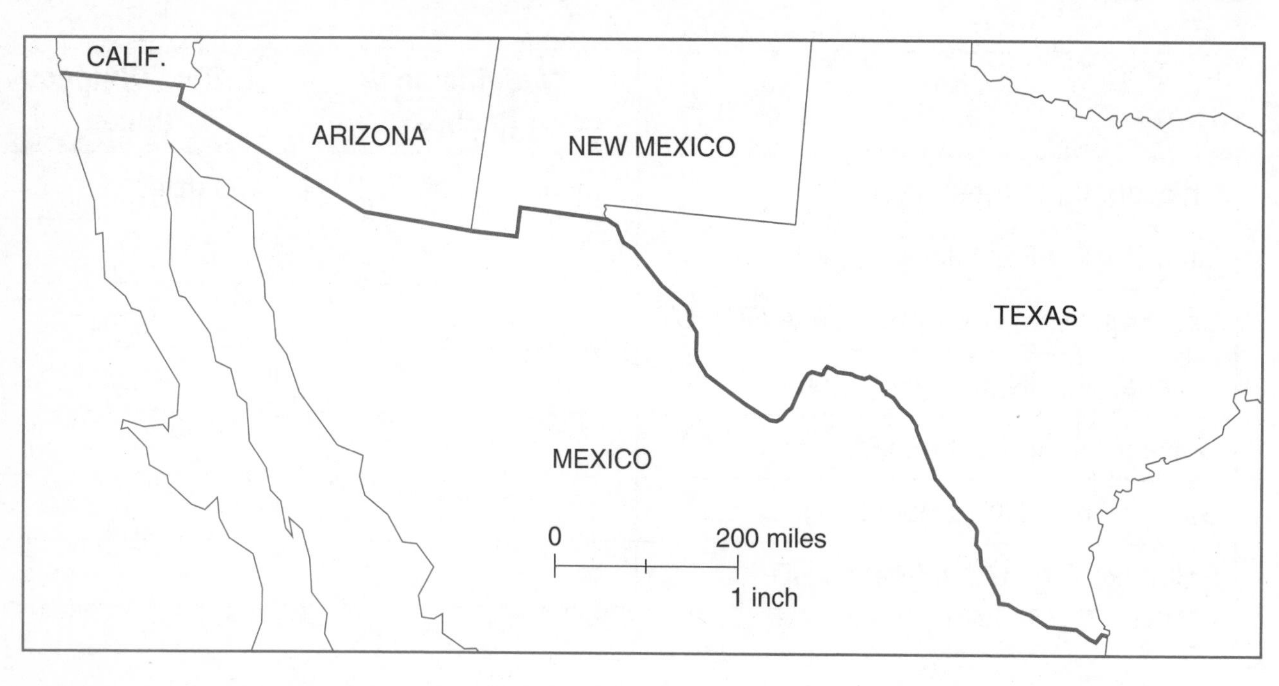

2. a. Estimate the lengths of the following rivers. Use the map on pages 344 and 345 of the *Student Reference Book.*

River	Length (miles)
Arkansas (CO, KS, OK, and AR)	
Missouri (MT, ND, SD, NE, IA, KS, and MO)	
Brazos (NM and TX)	
Chattahoochee (GA, AL, FL)	

b. Explain how you found the length of the Chattahoochee River.

______

______

______

______

Date Time

Classifying and Measuring Angles

Fill in the oval next to the correct answer for each angle.

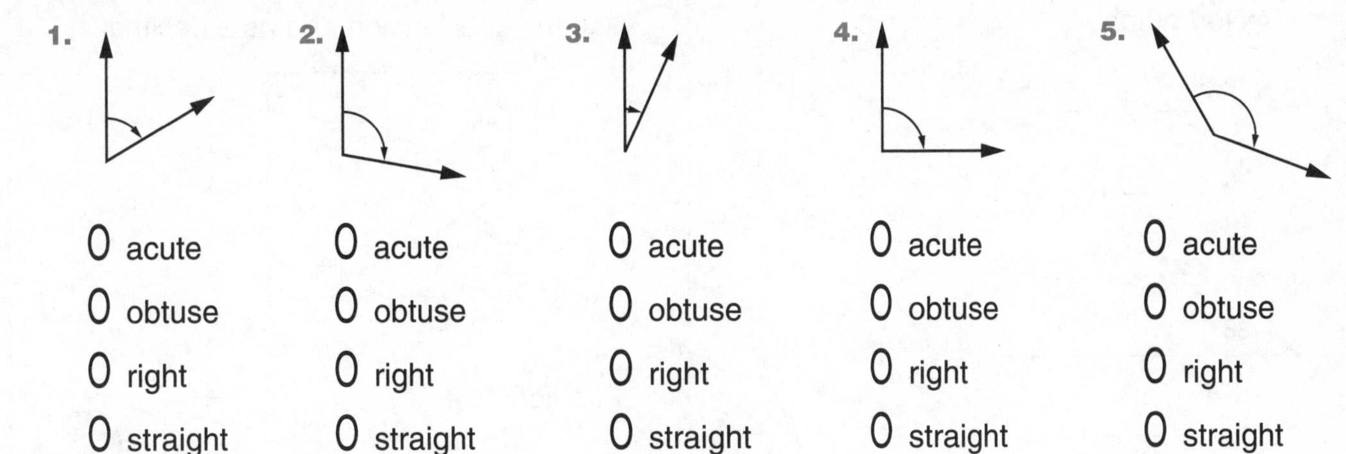

1.	2.	3.	4.	5.
O acute	O acute	O acute	O acute	O acute
O obtuse	O obtuse	O obtuse	O obtuse	O obtuse
O right	O right	O right	O right	O right
O straight	O straight	O straight	O straight	O straight

First, circle an estimate for the measure of each angle below. Then measure the angle.

6.

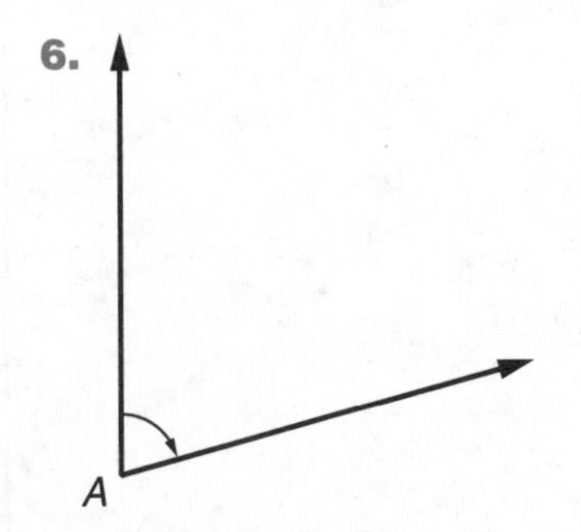

less than 90°
greater than 90°
equal to 90°

The measure of ∠A is about ________ °.

7.

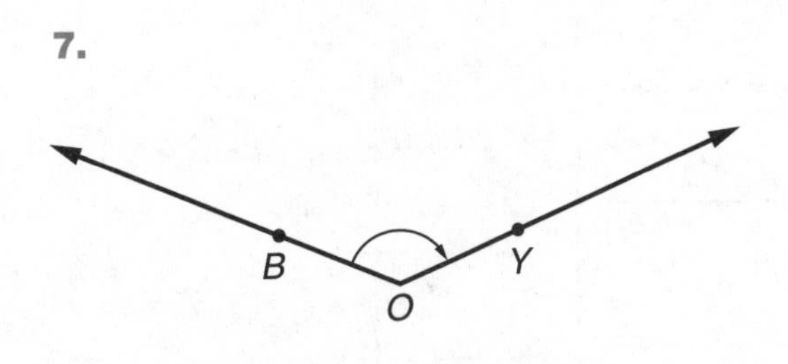

less than 90°
greater than 90°
equal to 90°

The measure of ∠BOY is about ________ °.

8.

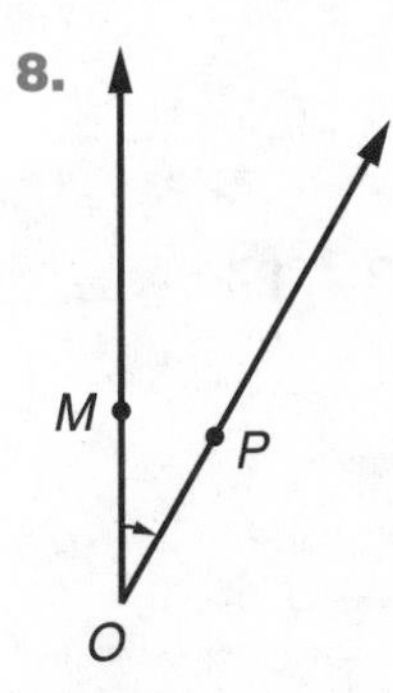

less than 90°
greater than 90°
equal to 90°

The measure of ∠MOP is about ________ °.

Use the figure to the right to answer Problems 9 and 10.

9. Name a pair of adjacent angles. ________ and ________

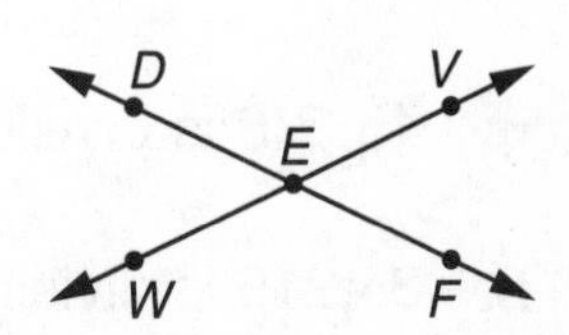

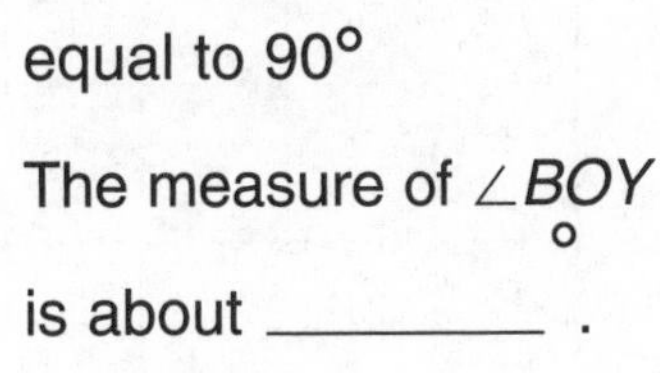

Challenge

10. The measure of ∠DEW is 50°. Without measuring, tell what the measure of ∠FEW is. ________ °

Date Time

Math Boxes 4.3

1. Write each fraction as a whole number or a mixed number.

a. $\frac{19}{5}$ ______________

b. $\frac{42}{8}$ ______________

c. $\frac{16}{6}$ ______________

d. $\frac{36}{12}$ ______________

e. $\frac{7}{4}$ ______________

2. Name the shaded part of the whole square as a fraction and as a decimal.

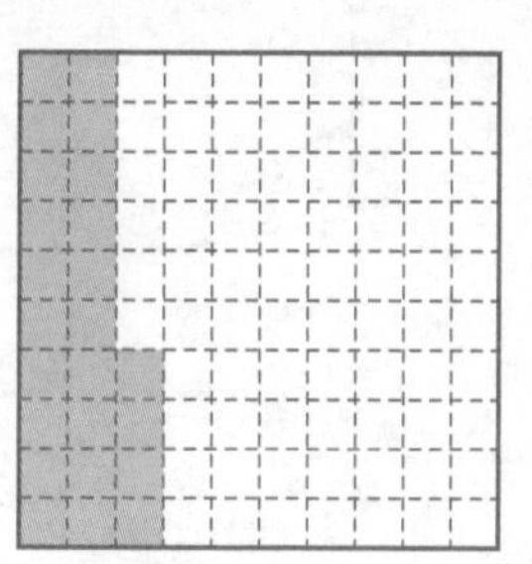

Fraction: ______________

Decimal: ______________

3. Multiply. Show your work.

a. $29 * 32 =$ ______________

b. $813 * 17 =$ ______________

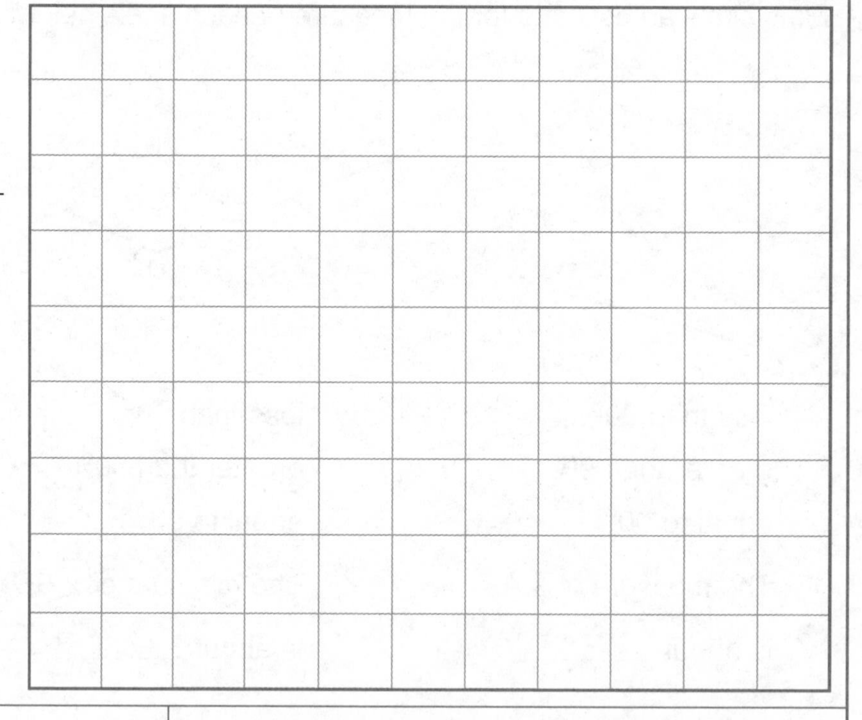

4. True or false. Write T or F.

a. 45,678 is divisible by 2. ________

b. 34,215 is divisible by 3. ________

c. 455 is divisible by 5. ________

d. 4,561 is divisible by 9. ________

5. Use your calculator to rename each of the following in standard notation.

a. $5^5 =$ ______________

b. $7^3 =$ ______________

c. $9^3 =$ ______________

d. $3^9 =$ ______________

Date Time

Estimate and Calculate Quotients

For each problem:

- Make a magnitude estimate of the quotient. Ask yourself, *Is the answer in the tenths, ones, tens, or hundreds?*
- Circle a box to show the magnitude of your estimate.
- Write a number sentence to show how you estimated.
- If there is a decimal point, ignore it. Divide the numbers.
- Use your magnitude estimate to place the decimal point in the final answer.

1. $3\overline{)36.6}$

0.1s	1s	10s	100s

How I estimated: ____________________

Answer: ________________

2. $4\overline{)9.48}$

0.1s	1s	10s	100s

How I estimated: ____________________

Answer: ________________

3. \$18.55 ÷ 7

0.1s	1s	10s	100s

How I estimated: ____________________

Answer: ________________

4. 7.842 ÷ 6

0.1s	1s	10s	100s

How I estimated: ____________________

Answer: ________________

5. 560.1 / 3

0.1s	1s	10s	100s

How I estimated: ____________________

Answer: ________________

6. 3.84 / 6

0.1s	1s	10s	100s

How I estimated: ____________________

Answer: ________________

Date Time

Math Boxes 4.4

1. Measure each line segment to the nearest centimeter.

   a. ____________

   ________ cm

   b. ____________________

   ________ cm

2. Each of my angles is greater than 90°. I have fewer than 6 sides.

   What shape am I?

   ____

   Use your Geometry Template to trace the shape below.

3. Estimate the answer to each multiplication problem.

   a. 45 * 19 = ____________

   b. 27 * 31 = ____________

   c. 52 * 87 = ____________

   d. 601 * 29 = ____________

   e. 398 * 42 = ____________

4. Round 16.354 to the nearest …

   a. ten. ____________

   b. tenth. ____________

   c. hundredth. ____________

5. Larry spent $4.82 on a notebook, $1.79 on paper to fill it, and $2.14 on a pen. How much did he spend in all?

   ____

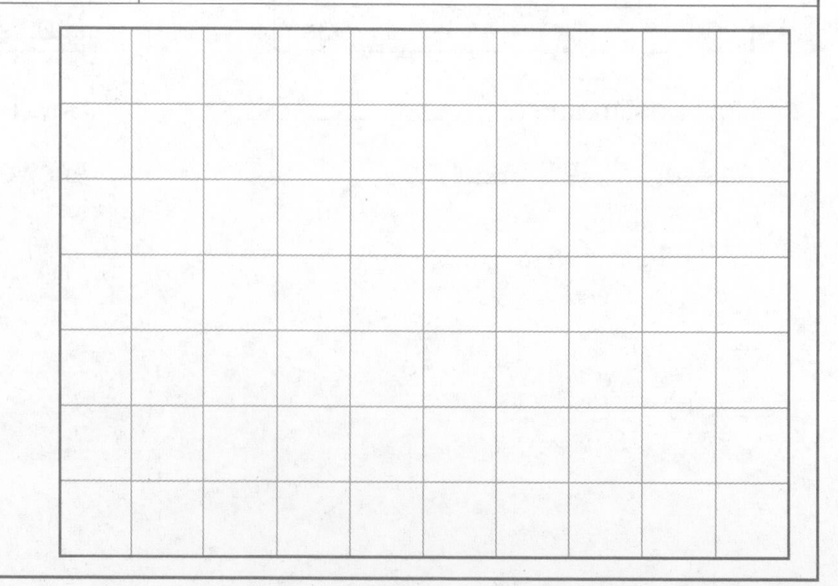

Date Time

Interpreting Remainders in Division Number Stories

For each number story:

- Draw a picture. Write a number sentence.
- Use a division algorithm to solve the problem.
- Tell what the remainder represents.
- Decide what to do about the remainder.

1. Compact discs are on sale for $9, including tax. How many can you buy with $30?

Picture:

Number sentence: ______________________

____

Solution: __________ compact discs

What does the remainder represent?

____

____

What did you do about the remainder? Circle the answer.

Ignored it.

Reported it as a fraction or decimal.

Rounded the answer up.

2. Rebecca and her three sisters bought their mother a bread machine for her birthday. The machine cost $219, including tax. The sisters split the bill evenly. How much did each sister contribute?

Picture:

Number sentence: ______________________

____

Solution: $__________

What does the remainder represent?

____

____

What did you do about the remainder? Circle the answer.

Ignored it.

Reported it as a fraction or decimal.

Rounded the answer up.

Date Time

Interpreting Remainders (cont.)

3. You are organizing a trip to a museum for 110 students, teachers, and parents. If each bus can seat 25 people, how many buses do you need?

Picture:

Number sentence: ______________________

____

Solution: __________ buses

What does the remainder represent?

____

____

____

What did you do about the remainder? Circle the answer.

Ignored it.

Reported it as a fraction or decimal.

Rounded the answer up.

Review: Magnitude Estimates and Division

4. $15\overline{)4,380}$

0.1s	1s	10s	100s

How I estimated: ______________________

Answer: ______________

5. $3\overline{)70.5}$

0.1s	1s	10s	100s

How I estimated: ______________________

Answer: ______________

6. 82.8 / 12

0.1s	1s	10s	100s

How I estimated: ______________________

Answer: ______________

Challenge

7. 3.75 / 25

0.1s	1s	10s	100s

How I estimated: ______________________

Answer: ______________

Date Time

Math Boxes 4.5

1. Measure the length and width of each of the following objects to the nearest half inch.

a. journal cover length ________ in. width ________ in.

b. desktop length ________ in. width ________ in.

c. index card length ________ in. width ________ in.

d. ________________ length ________ in. width ________ in.
(your choice)

2. Do the following multiplication problems mentally.

a. 79 * 8 = ________

b. ________ = 299 * 4

c. ________ = 25 * 99

d. 69 * 7 = ________

e. 499 * 6 = ________

3. Write the following numbers in order from greatest to least.

0.38 0.308 3.08 3.38 0.038

____

____

____

____

____

4. Measure each angle to the nearest degree.

a.

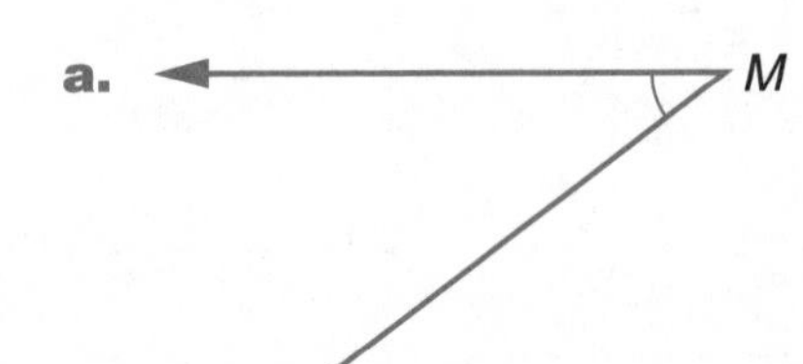

The measure of $\angle M$ is about ________ °.

b.

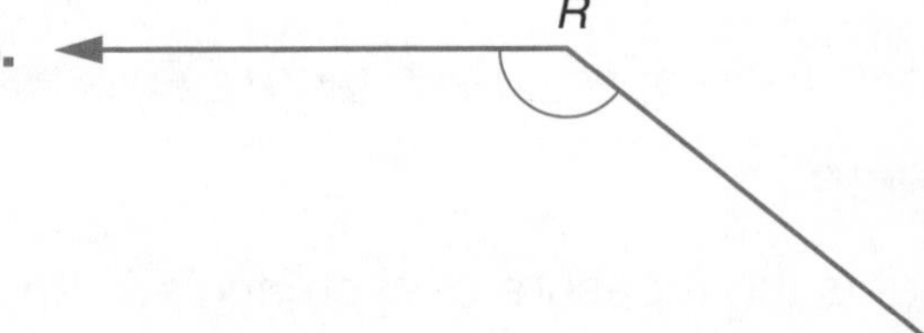

The measure of $\angle R$ is about ________ °.

Date Time

Triangle and Polygon Review

Fill in the oval next to the correct answer for each triangle.

1. 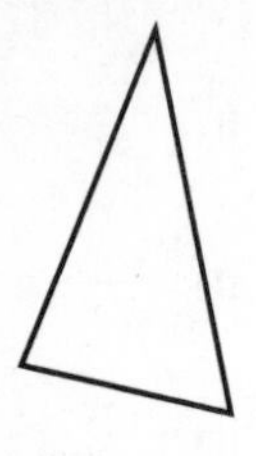

O equilateral
O isosceles
O scalene

2. 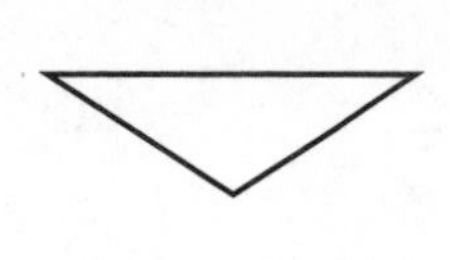

O equilateral
O isosceles
O right

3. 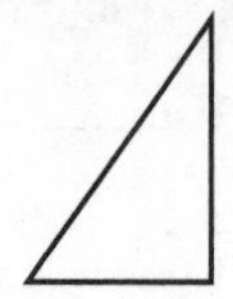

O equilateral
O isosceles
O right

4. 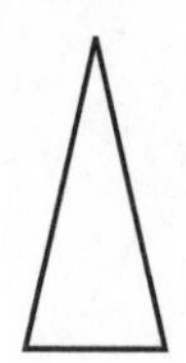

O equilateral
O isosceles
O scalene

5. 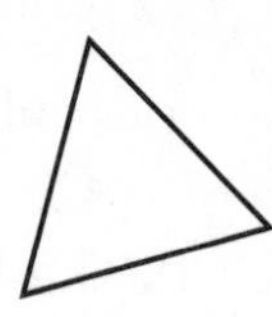

O equilateral
O right
O scalene

6. Marlene drew four shapes—an isosceles triangle, a pentagon, a trapezoid, and a rectangle. She covered up most of each figure as shown below. Write the name below each figure. Draw the rest of the figure.

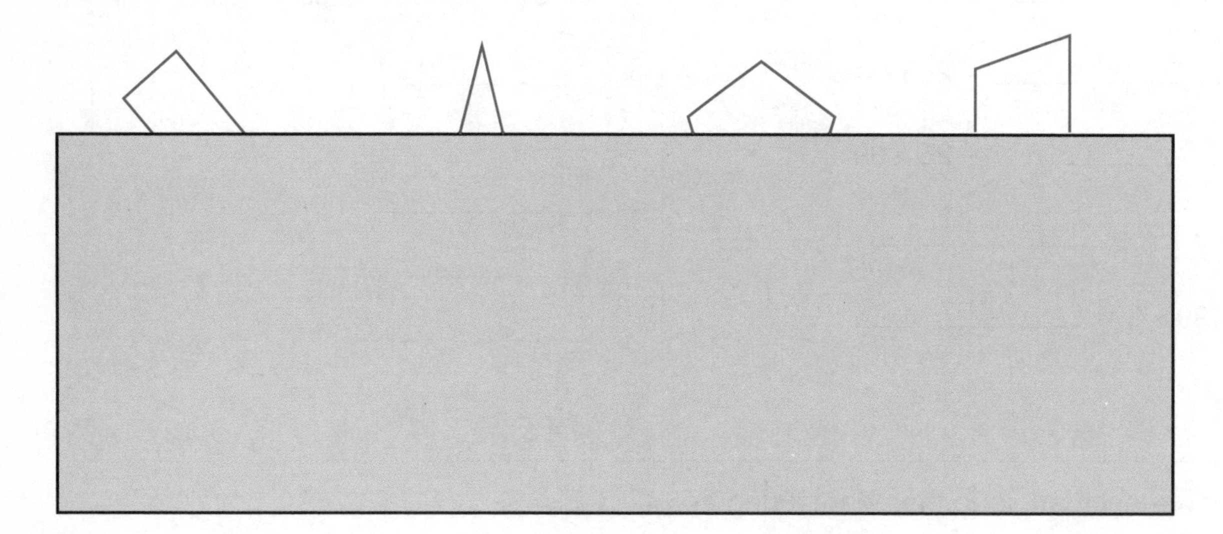

Challenge

7. What is the measure of each angle in an equilateral triangle? ________

Explain how you know. ____

____

____

____

Date Time

Math Boxes 4.6

1. Write each fraction as a whole number or a mixed number.

   a. $\frac{24}{8}$ ________

   b. $\frac{18}{5}$ ________

   c. $\frac{21}{6}$ ________

   d. $\frac{15}{4}$ ________

   e. $\frac{11}{3}$ ________

2. Name the shaded part of the whole square as a fraction and as a decimal.

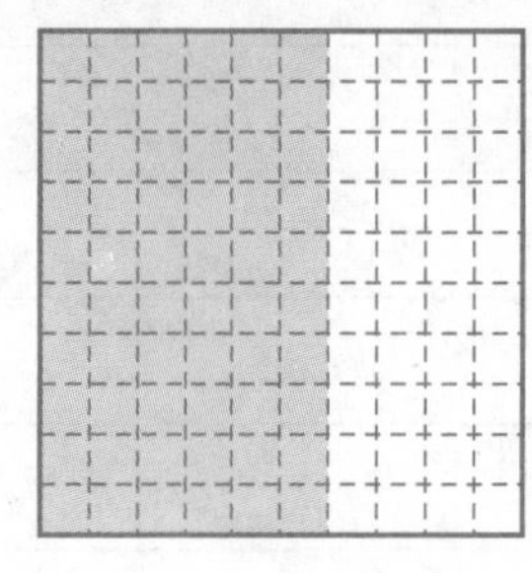

Fraction: ____________

Decimal: ____________

3. Multiply. Show your work.

   a. $41 * 69 =$ ____________

   b. $803 * 37 =$ ____________

4. True or false? Write T or F.

   a. 5,278 is divisible by 3. ________

   b. 79,002 is divisible by 6. ________

   c. 86,076 is divisible by 9. ________

   d. 908,321 is divisible by 2. ________

5. Using your calculator, find the square root of each of the following numbers.

   a. 361 ________

   b. 2,704 ________

   c. 8,649 ________

   d. 4,356 ________

Date Time

Time to Reflect

1. Tell why you think it is important to be able to divide numbers. For what kind of problems do you need to use division?

2. Tell what part of this unit was the most difficult for you and why. Describe what you did to overcome any difficulties you had.

Math Boxes 4.7

1. Measure each line segment to the nearest quarter-inch.

a. 

_________ in.

b. 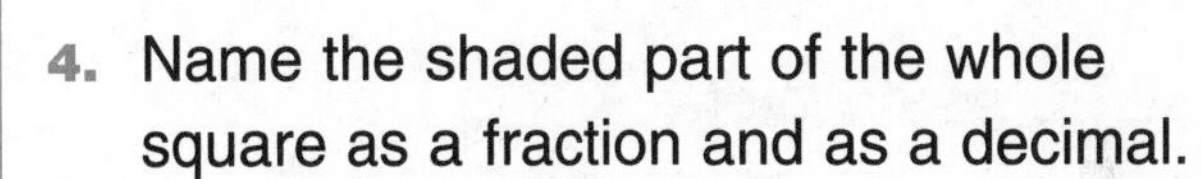

_________ in.

2. Write each fraction as a whole number or a mixed number.

a. $\frac{17}{4}$ _________

b. $\frac{24}{3}$ _________

c. $\frac{5}{2}$ _________

d. $\frac{9}{8}$ _________

e. $\frac{32}{5}$ _________

3. Show $\frac{2}{5}$ in at least two different ways.

4. Name the shaded part of the whole square as a fraction and as a decimal.

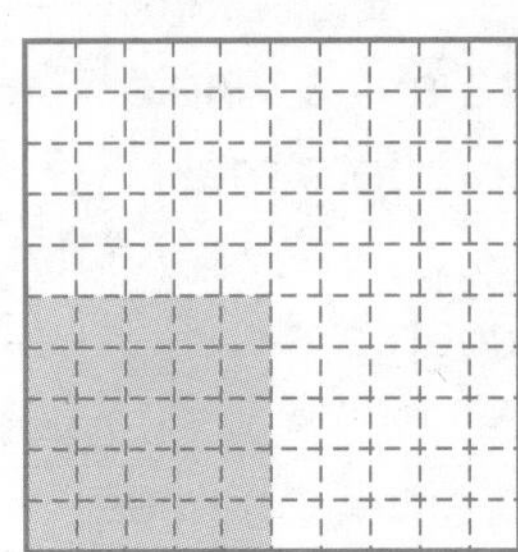

Fraction: _________

Decimal: _________

5. Write each mixed number as an improper fraction.

a. $1\frac{3}{4}$ _________

b. $3\frac{1}{2}$ _________

c. $2\frac{7}{8}$ _________

d. $4\frac{9}{5}$ _________

e. $6\frac{1}{3}$ _________

6. Measure the dimensions of your calculator to the nearest $\frac{1}{4}$ inch. Record your measurements on the drawing below.

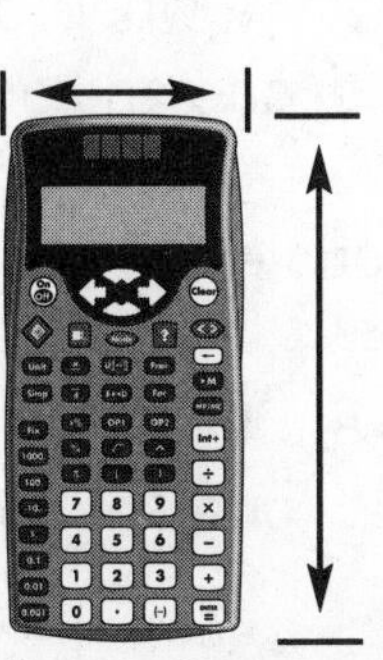

Date Time

Parts and Wholes

Work with a partner. Use counters to help you solve these problems.

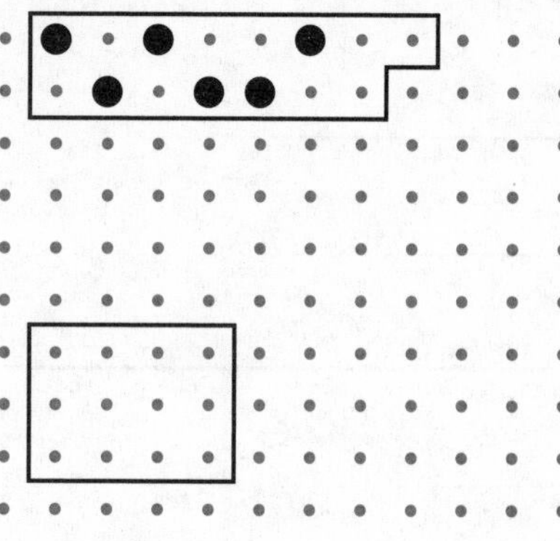

1. This set has 15 counters.
   What fraction of the set is black?

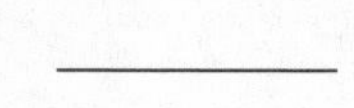

2. If 12 counters are the whole set,
   what fraction of the set is 8 counters?

   ____

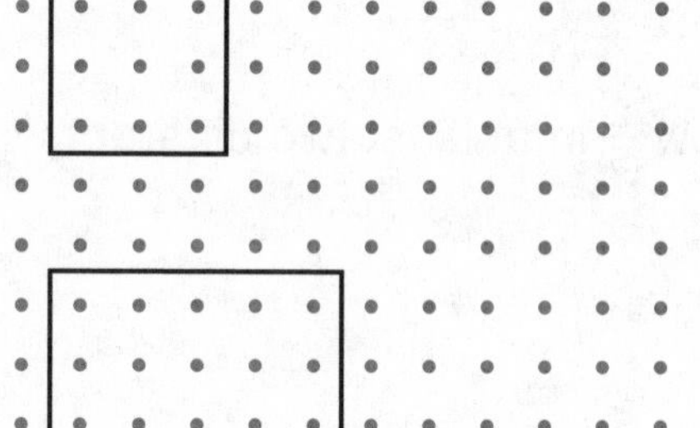

3. If 12 counters are the whole set,
   how many counters is $\frac{1}{4}$ of a set?

 counters

4. If 20 counters are a whole,
   how many counters make $\frac{4}{5}$?

 counters

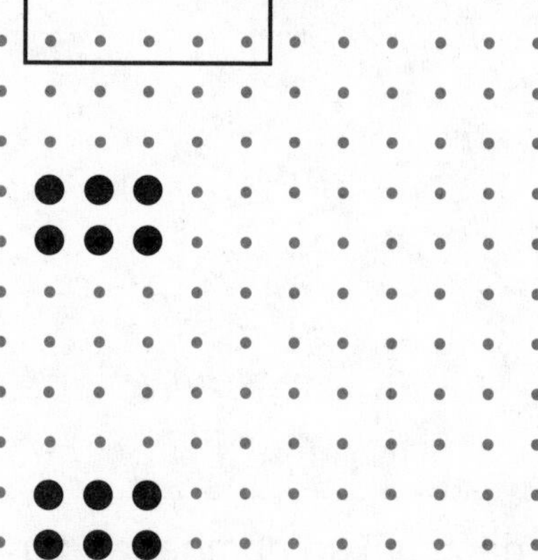

5. If 6 counters are $\frac{1}{2}$ of a set,
   how big is the set?

   ________ counters

6. If 12 counters are $\frac{3}{4}$ of a set, how
   many counters are in the whole set?

   ________ counters

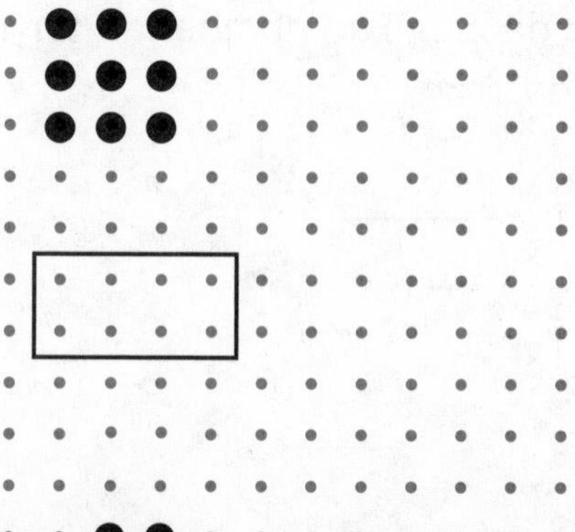

7. If 8 counters are a whole set, how many
   counters are in one and one-half sets?

   ________ counters

8. If 6 counters are two-thirds of a set, how many
   counters are in one and two-thirds sets?

   ________ counters

Date Time

Finding Fractions of a Whole

1. In a school election, 141 fifth graders voted. One-third voted for Shira and two-thirds voted for Bree.

   a. How many votes did Shira get? ________

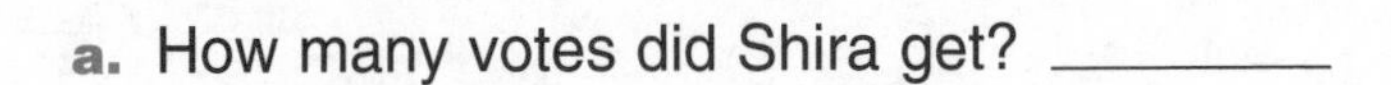

   b. How many votes did Bree get? ________

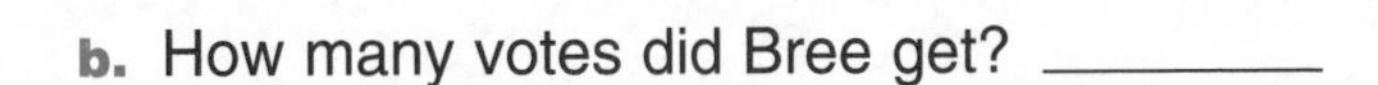

2. Bob, Liz, and Eli drove from Chicago to Denver.
   Bob drove $\frac{1}{10}$ of the distance.
   Liz drove $\frac{4}{10}$ of the distance.
   Eli drove $\frac{1}{2}$ of the distance.
   How many miles did each person drive?
   Check to make sure that the total is 1,050 miles.

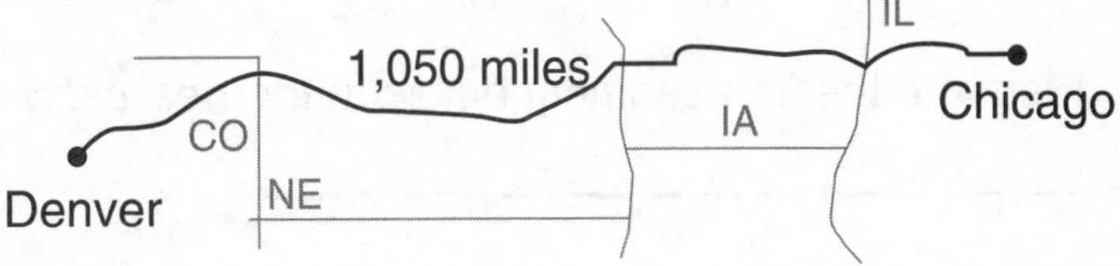

   a. Bob: ________ miles b. Liz: ________ miles c. Eli: ________ miles

3. Carlos and Rick paid \$8.75 for a present. Carlos paid $\frac{2}{5}$ of the total amount and Rick paid $\frac{3}{5}$ of the total.

   a. How much did Carlos pay? ____________

   b. How much did Rick pay? ____________

4. A pizza costs \$12.00, including tax. Scott paid $\frac{1}{4}$ of the total cost. Trung paid $\frac{1}{3}$ of the total cost. Pritish paid $\frac{1}{6}$. Bill paid the rest. How much did each person pay?

   a. Scott: \$______ b. Trung: \$______ c. Pritish: \$______ d. Bill: \$______

5. If 60 counters are the whole, how many counters make two-thirds?

   ________ counters

6. If 75 counters are $\frac{3}{4}$ of a set, how many counters are in the whole set?

   ________ counters

7. If 15 counters are a whole, how many counters make three-fifths?

   ________ counters

Date Time

Reading a Ruler

1. Use your ruler. Measure each line segment below to the *nearest half-inch.*

a. ____

______ inches

b. ______________________ **c.** _________________

______ inches ______ inches

2. Measure the line segment below to the *nearest quarter inch.*

____

______ inches

3. Compare each pair of lengths below. First, use your ruler to mark the line segments. Then write $<$, $=$, or $>$.

$<$ means *is less than*
$=$ means *equals*
$>$ means *is greater than*

a. $1\frac{1}{4}$ inches ______ $1\frac{1}{8}$ inches

b. $2\frac{3}{4}$ inches ______ 3 inches

c. $2\frac{2}{4}$ inches ______ $2\frac{1}{2}$ inches

d. $2\frac{3}{4}$ inches ______ $1\frac{3}{4}$ inches

4. a. Mark a line segment that is $2\frac{4}{8}$ inches long.

b. How many half-inches long is it? ______ half-inches

5. a. Mark a line segment that is 5 quarter-inches long.

b. This is the same as (circle one) $1\frac{1}{4}$ inches. $1\frac{2}{4}$ inches. $1\frac{3}{4}$ inches.

Date Time

Math Boxes 5.1

1. Write five fractions that are equivalent to $\frac{1}{2}$.

____

____

____

____

____

SRB 59–61

2. Round each number to the nearest ten-thousand.

a. 1,308,799 _______________

b. 621,499 _______________

c. 8,003,291 _______________

d. 158,005 _______________

e. 2,226,095 _______________

3. Multiply or divide.
Show your work.

a. 58 * 73 = _______________

b. 793 ÷ 8 = _______________

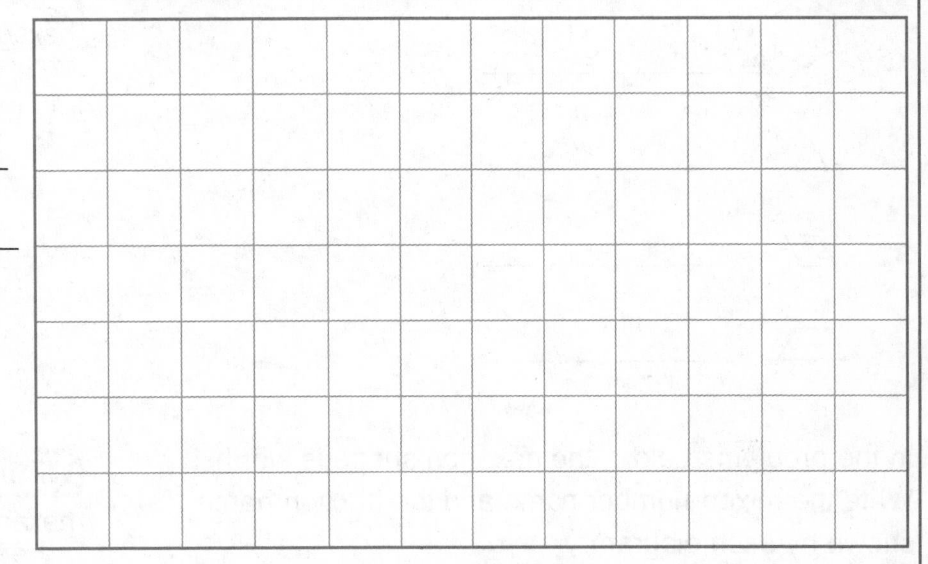

4. a. Draw two lines that meet at right angles.

b. What is the size of each angle?

____

5. Trace a scalene triangle from your Geometry Template.

SRB 134 152 153

Date Time

Mixed Numbers: Part 1

Fractions greater than 1 can be written in several different ways.

Example If a circle  is worth 1, what is 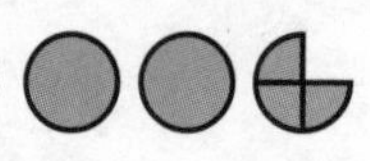 worth?

The mixed-number name is $2\frac{3}{4}$. ($2\frac{3}{4}$ means $2 + \frac{3}{4}$.)

The fraction name is $\frac{11}{4}$. (Think quarters: ⊕⊕◔.)

So $2\frac{3}{4}$, $2 + \frac{3}{4}$, and $\frac{11}{4}$ are just different names for the same number.

In the problems below, the hexagon shape is worth 1.

Whole
hexagon

1. ⬡ = ____________

2. △ = ____________

3. ▱ = ____________

4. ⏢ = ____________

In the problems below, the hexagon shape is worth 1. Write the mixed-number name and the fraction name shown by each diagram.

Whole
hexagon

5. 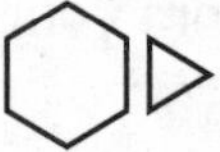 Mixed number = Fraction =

6. 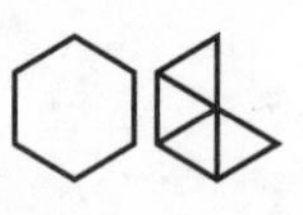 Mixed number = Fraction =

7. 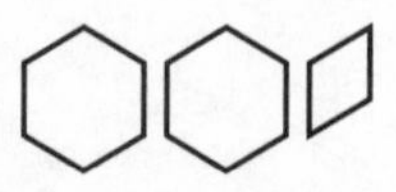 Mixed number = Fraction =

8. 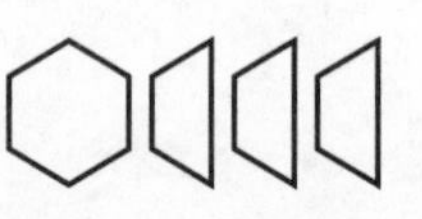 Mixed number = Fraction =

9. 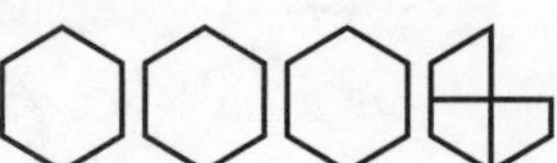 Mixed number = Fraction =

Date Time

Mixed Numbers: Part 2

For Problems 1–5, each triangle block is worth $\frac{1}{4}$.

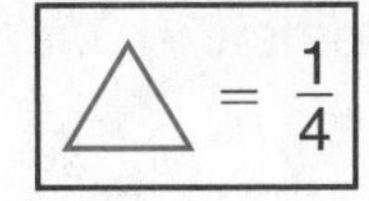

Use your , , and pattern blocks to solve these problems.

1. Cover a rhombus block with triangle blocks. A rhombus is worth ________.

2. Cover a trapezoid block with triangle blocks. A trapezoid is worth ________.

3. Arrange your blocks to make a shape worth 1. Trace the outline of each block that is part of your shape, or use your Geometry Template. Label each part with a fraction.

4. Arrange your blocks to make a shape that is worth $2\frac{1}{2}$. Trace the outline of each block that is part of your shape, or use your Geometry Template. Label each part with a fraction.

5. Use your blocks to cover this shape.

   Trace the outline of each block and label each part with a fraction.

   How much is the shape worth?

   ____

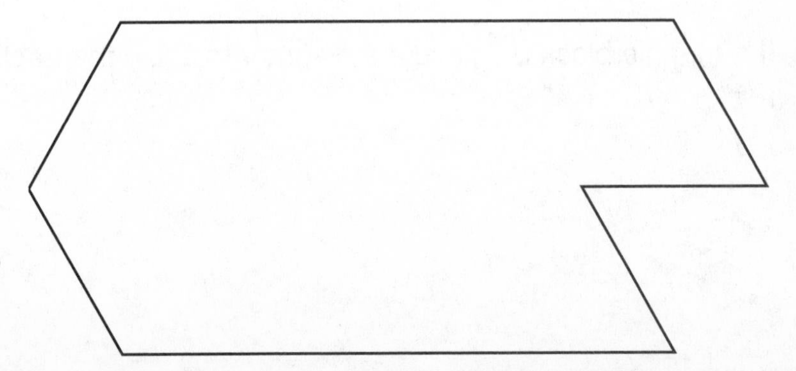

Date Time

Mixed Numbers: Part 2 (cont.)

For Problems 6–10, each triangle block is worth $\frac{1}{2}$.

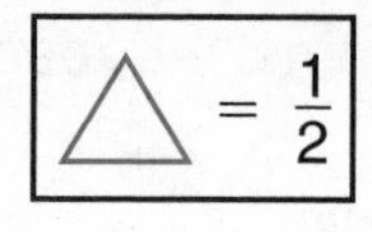

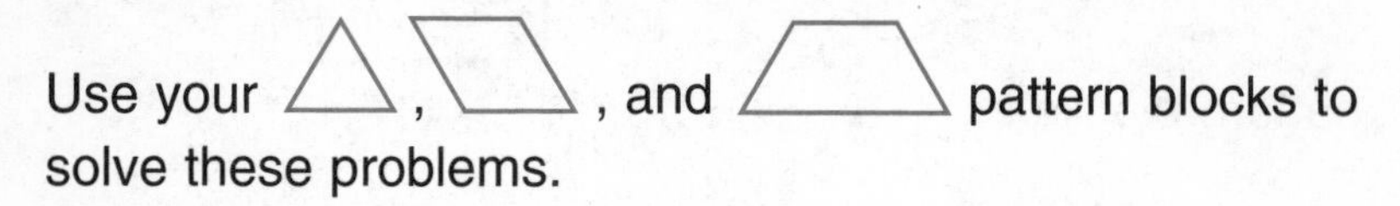

Use your △, ▱, and ⏢ pattern blocks to solve these problems.

6. What shape is worth ONE? ______________________

7. A rhombus is worth ______________.

8. A trapezoid is worth ______________.

9. Arrange your blocks to make a shape that is worth $3\frac{1}{2}$. Trace the outline of each block that is part of your shape, or use your Geometry Template. Label each part with a fraction.

10. Use your blocks to cover the shape below. Trace the outline of each block. Label each part with a fraction.

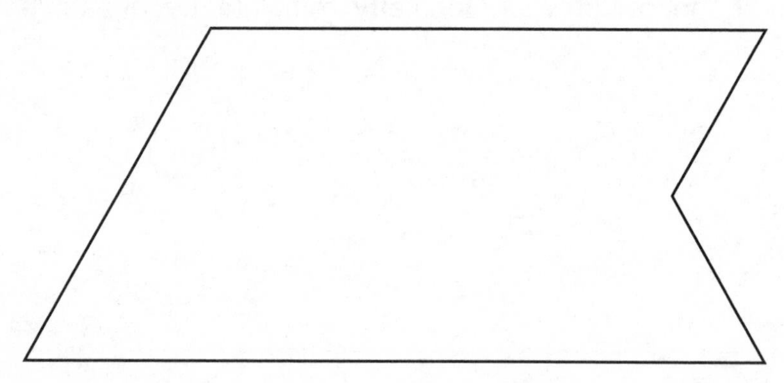

How much is the shape worth? ______________

11. If a triangle block is $\frac{1}{4}$, make a diagram to show the fraction $\frac{15}{4}$.

△ = $\frac{1}{4}$

Write $\frac{15}{4}$ as a mixed number. $\frac{15}{4}$ = ________.

Date Time

Fractions on a Ruler

1. Find and mark each of these lengths on the ruler below. Write the letter above the mark. Letters *A* and *B* are done for you.

A: 5"	*B*: $\frac{1}{2}$"	*C*: $3\frac{1}{2}$"	*D*: $2\frac{1}{2}$"
E: $4\frac{3}{4}$"	*F*: $\frac{1}{4}$"	*G*: $4\frac{1}{8}$"	*H*: $1\frac{7}{8}$"
I: $1\frac{3}{8}$"	*J*: $\frac{15}{16}$"	*K*: $3\frac{1}{16}$"	*L*: $5\frac{9}{16}$"

2. On the ruler above, how many fractions are shown between 0 and 1? Explain.

____

____

3. Grace was supposed to mark $\frac{1}{2}$ on a number line. This is what she did.

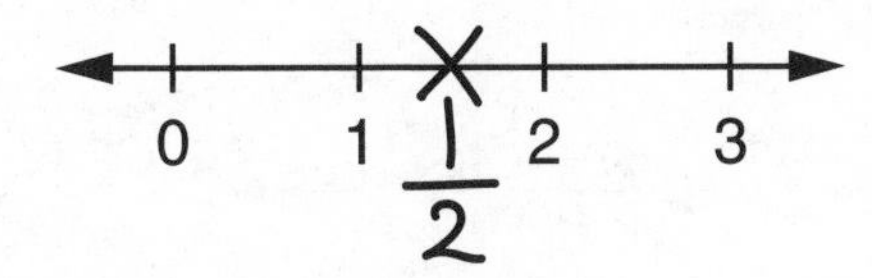

Explain Grace's mistake. ____

____

4. Rocco said this stick is $4\frac{3}{16}$ inches long.

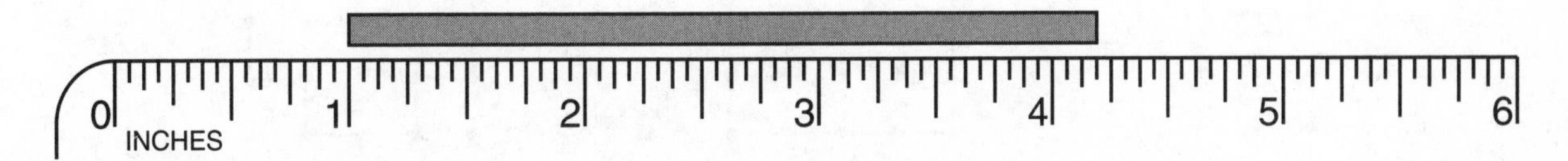

Explain his mistake. ____

____

Date Time

Math Boxes 5.2

1. Find the landmarks for this set of numbers: 273, 280, 298, 254, 328, 269, 317, 280, 309

   a. Maximum: ______________

   b. Minimum: ______________

   c. Range: ______________

   d. Median: ______________

2. Solve mentally.

   a. 99 * 37 = ______________

   b. 15 * 399 = ______________

   c. 20 * 599 = ______________

   d. ______________ = 899 * 30

   e. ______________ = 68 * 99

3. Write a number story for the number model 837 / 7 = *m*. Then solve it.

____

____

____

____

____

____

Answer: ______________

4. What is the perimeter of each shape?

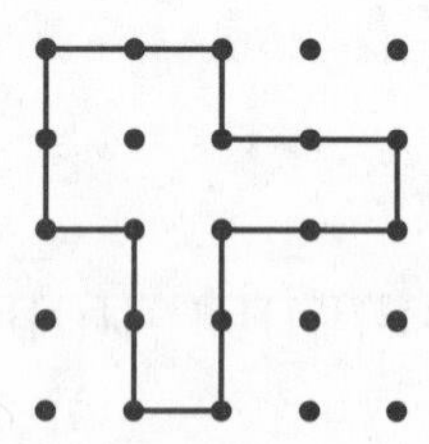

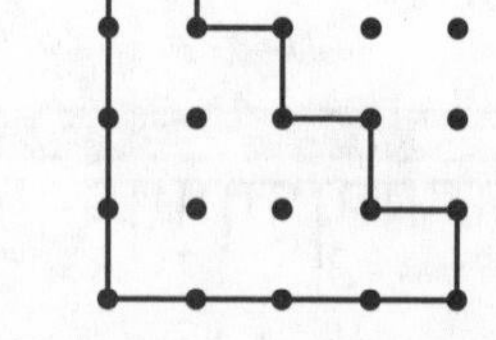

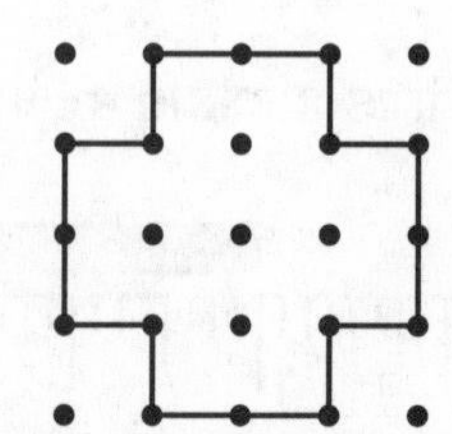

P = ________ units　*P* = ________ units　*P* = ________ units

Date Time

Comparing and Ordering Fractions

Math Message

Decide for each of these measurements whether it is closest to 0, $\frac{1}{2}$, or 1 inch. Circle the measurement it is closest to.

1. $\frac{1}{8}$ inch is closest to 0 inches. $\frac{1}{2}$ inch. 1 inch.

2. $\frac{15}{16}$ inch is closest to 0 inches. $\frac{1}{2}$ inch. 1 inch.

3. $\frac{5}{8}$ inch is closest to 0 inches. $\frac{1}{2}$ inch. 1 inch.

4. $\frac{3}{8}$ inch is closest to 0 inches. $\frac{1}{2}$ inch. 1 inch.

5. Rewrite the following fractions in order from least to greatest.

   $\frac{1}{8}, \frac{15}{16}, \frac{5}{8}, \frac{3}{8}$ ______, ______, ______, ______

Ordering Fractions

For each problem below, write the fractions in order from least to greatest.

6. $\frac{6}{8}, \frac{3}{8}, \frac{5}{8}, \frac{8}{8}$ ______, ______, ______, ______

7. $\frac{2}{7}, \frac{2}{9}, \frac{2}{5}, \frac{2}{12}$ ______, ______, ______, ______

8. $\frac{2}{3}, \frac{1}{4}, \frac{1}{3}, \frac{3}{4}$ ______, ______, ______, ______

9. $\frac{3}{5}, \frac{4}{10}, \frac{9}{20}, \frac{1}{25}$ ______, ______, ______, ______

10. $\frac{3}{7}, \frac{1}{10}, \frac{7}{8}, \frac{5}{7}$ ______, ______, ______, ______

11. $\frac{5}{9}, \frac{2}{5}, \frac{1}{6}, \frac{9}{10}$ ______, ______, ______, ______

12. $\frac{4}{8}, \frac{4}{7}, \frac{3}{5}, \frac{4}{9}$ ______, ______, ______, ______

Date Time

Fraction-Stick Chart

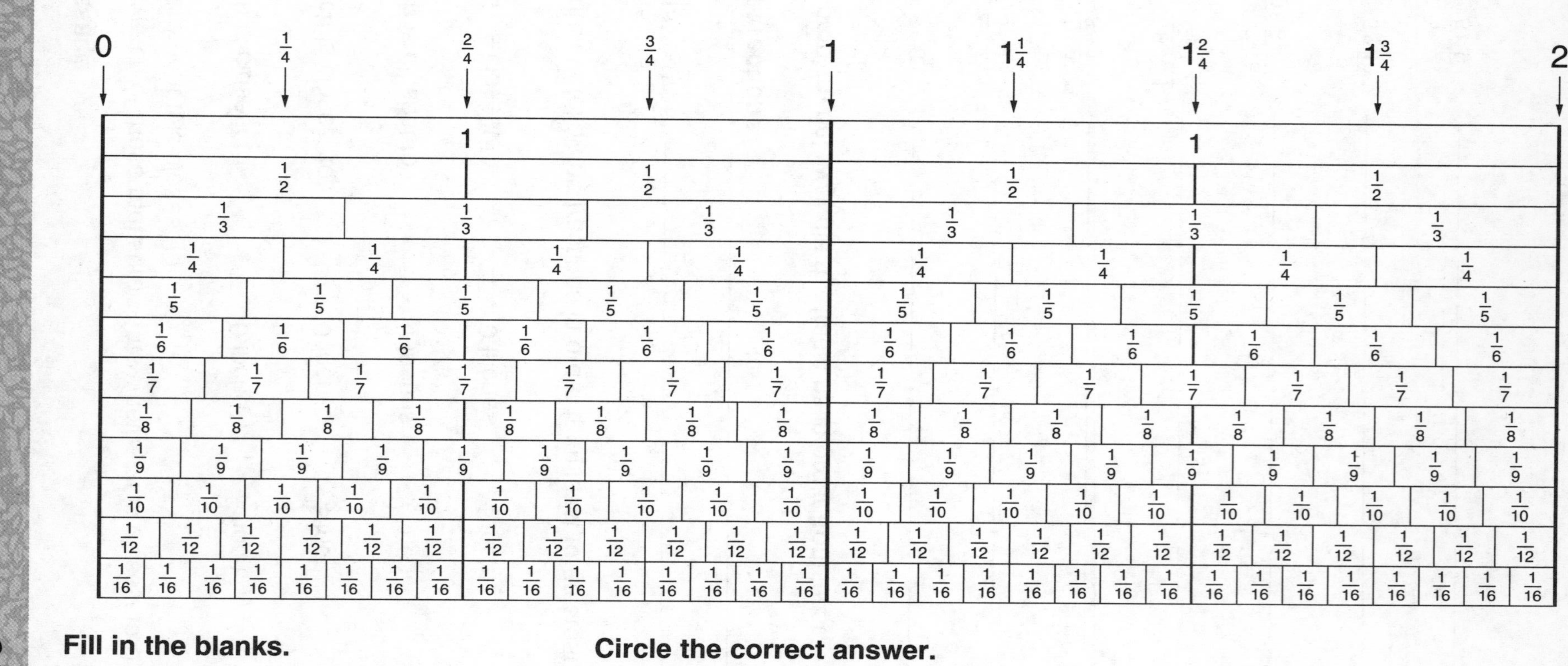

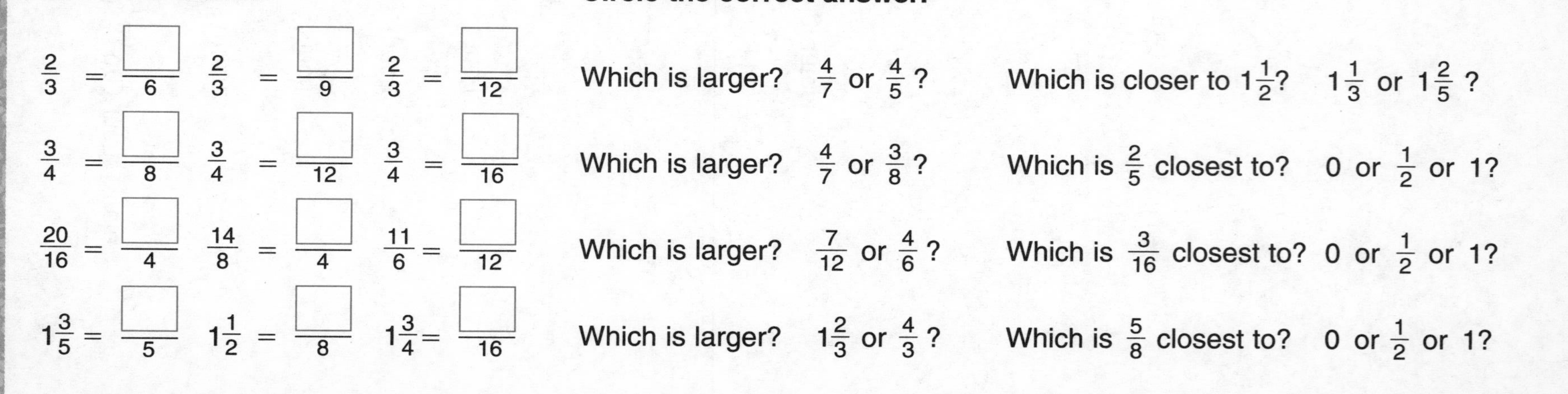

Fill in the blanks.

$\frac{2}{3} = \frac{\square}{6}$ $\frac{2}{3} = \frac{\square}{9}$ $\frac{2}{3} = \frac{\square}{12}$

$\frac{3}{4} = \frac{\square}{8}$ $\frac{3}{4} = \frac{\square}{12}$ $\frac{3}{4} = \frac{\square}{16}$

$\frac{20}{16} = \frac{\square}{4}$ $\frac{14}{8} = \frac{\square}{4}$ $\frac{11}{6} = \frac{\square}{12}$

$1\frac{3}{5} = \frac{\square}{5}$ $1\frac{1}{2} = \frac{\square}{8}$ $1\frac{3}{4} = \frac{\square}{16}$

Circle the correct answer.

Which is larger? $\frac{4}{7}$ or $\frac{4}{5}$?

Which is larger? $\frac{4}{7}$ or $\frac{3}{8}$?

Which is larger? $\frac{7}{12}$ or $\frac{4}{6}$?

Which is larger? $1\frac{2}{3}$ or $\frac{4}{3}$?

Which is closer to $1\frac{1}{2}$? $1\frac{1}{3}$ or $1\frac{2}{5}$?

Which is $\frac{2}{5}$ closest to? 0 or $\frac{1}{2}$ or 1?

Which is $\frac{3}{16}$ closest to? 0 or $\frac{1}{2}$ or 1?

Which is $\frac{5}{8}$ closest to? 0 or $\frac{1}{2}$ or 1?

Date Time

Fraction-Stick Pieces

A whole stick is worth 1. = 1

= 2 halves

= 4 quarters

= 8 eighths

= 16 sixteenths

1. Use the fraction sticks to find equivalent fractions.

a. $\frac{1}{8} = \frac{\square}{16}$ **b.** $\frac{\square}{8} = \frac{12}{16} = \frac{\square}{4}$ **c.** $\frac{\square}{8} = \frac{3}{4} = \frac{\square}{16}$

d. $\frac{1}{2} = \frac{\square}{4} = \frac{\square}{8} = \frac{\square}{16}$ **e.** $\frac{\square}{2} = \frac{4}{4} = \frac{\square}{8} = \frac{\square}{16}$

2. Use the fraction sticks to add fractions with the same denominator.

Example $\frac{1}{8} + \frac{2}{8} =$ 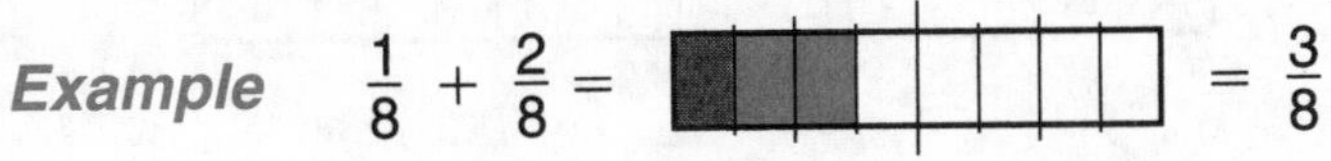 $= \frac{3}{8}$

a. $\frac{2}{4} + \frac{1}{4} =$ = ______

b. $\frac{3}{16} + \frac{9}{16} =$ = ______

c. $\frac{1}{16} + \frac{5}{16} + \frac{8}{16} =$ = ______

3. Use the fraction sticks to add fractions having different denominators.

a. $\frac{1}{2} + \frac{1}{4} =$ = ______

b. $\frac{1}{2} + \frac{3}{8} =$ = ______

c. $\frac{5}{8} + \frac{1}{4} =$ = ______

d. $\frac{1}{4} + \frac{7}{8} + \frac{2}{16} =$ = ______

Date Time

Fraction Number Stories

Shade the fraction sticks to help you solve these fraction number stories. Write a number model for each story.

1. Chris made pizza dough with $\frac{5}{8}$ cup of white flour and $\frac{1}{4}$ cup of whole wheat flour.

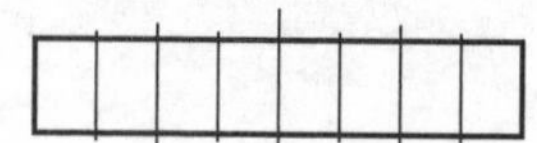

   a. How much flour did he use in all? _______ cup

   b. Number model: _______________________

2. Sheryl's puppy weighed $1\frac{1}{2}$ pounds when it was born. After two weeks, the puppy had gained $\frac{3}{8}$ pounds.

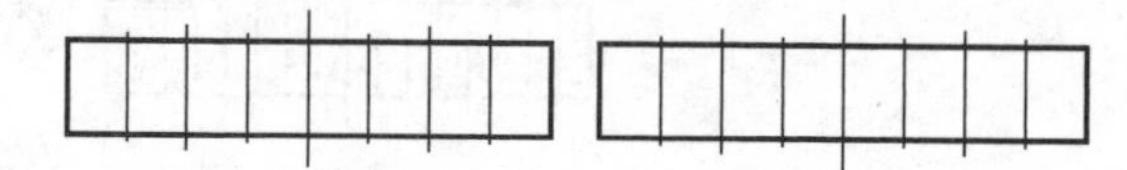

   a. How much did the puppy weigh after two weeks? _______ pounds

   b. Number model: _______________________

3. Shade the fraction sticks to solve the number model. Then write a fraction number story that fits the number model.

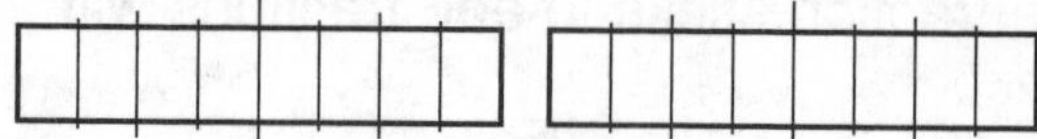

   a. $\frac{3}{4} + \frac{5}{8} =$ _______

   b. Number story: ____

   ____

4. Make up your own fraction number story. Draw and shade fraction sticks to solve it. Write a number model for your story.

   a. Number story: ____

   ____

   b. Solution: _______

   c. Number model: _______________________

Date Time

Math Boxes 5.3

1. Write five fractions that are equivalent to $\frac{3}{4}$.

____

____

____

____

____

2. Round each number to the nearest thousand.

a. 43,802 _______________

b. 904,873 _______________

c. 1,380,021 _______________

d. 5,067 _______________

e. 20,503 _______________

3. Multiply or divide. Show your work.

a. _______________ = 38 ∗ 47

b. _______________ → 857 / 6

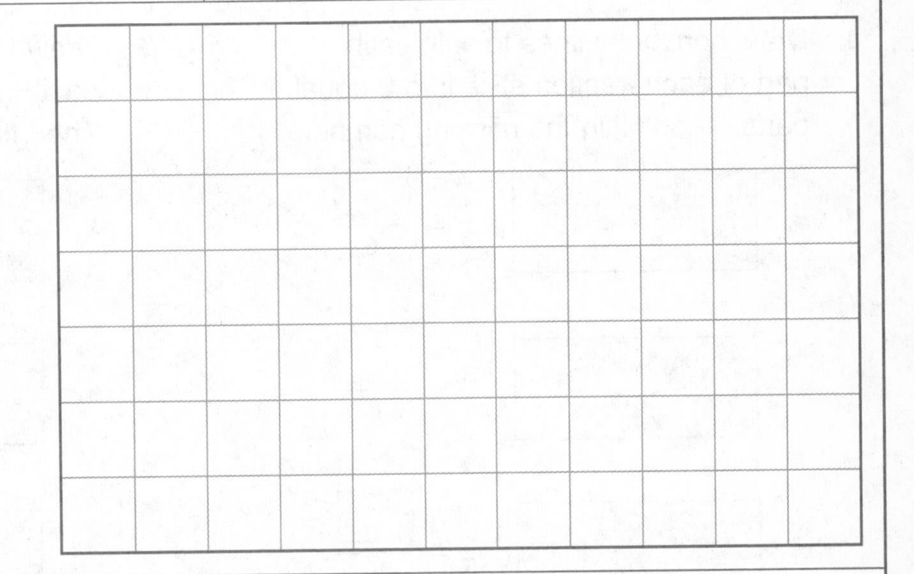

4. **a.** Draw a quadrangle with two pairs of parallel sides.

b. What kind of quadrangle is this?

____

5. Trace the equilateral triangle from your Geometry Template.

Date Time

Finding Equivalent Fractions by Splitting Fraction Sticks

Here is a way to get equivalent fractions. Start with a fraction stick that shows 3 out of 7 parts ($\frac{3}{7}$) shaded.

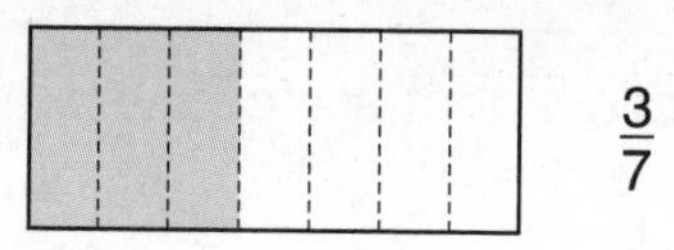

$\frac{3}{7}$

Draw a horizontal line to split each part of the stick into 2 equal parts. Now 6 out of 14 parts ($\frac{6}{14}$) are shaded. So $\frac{3}{7} = \frac{6}{14}$.

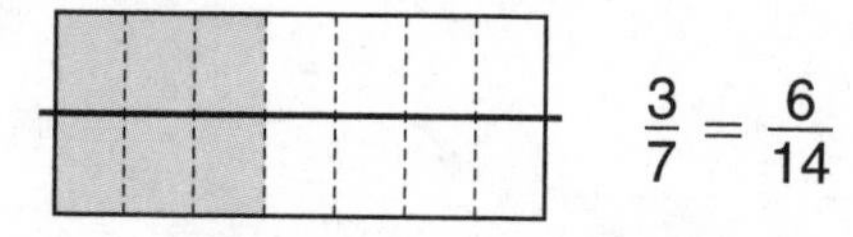

$\frac{3}{7} = \frac{6}{14}$

If each part of the original fraction stick is split into 3 equal parts, 9 out of 21 parts ($\frac{9}{21}$) are shaded. So $\frac{3}{7} = \frac{9}{21}$.

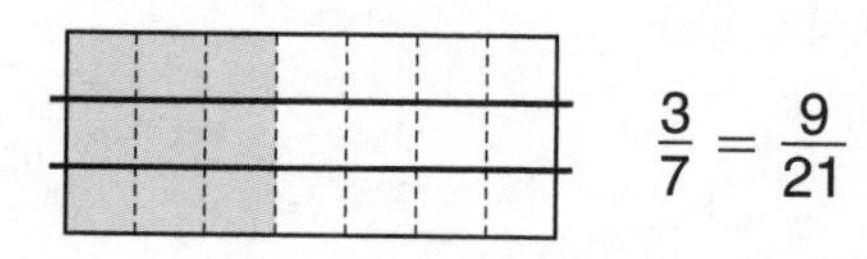

$\frac{3}{7} = \frac{9}{21}$

1. Draw horizontal lines to split each part of each fraction stick into 2 equal parts. Then fill in the missing numbers.

a. 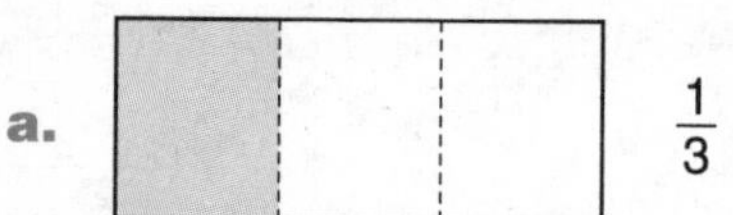 $\frac{1}{3} = \frac{\square}{6}$

b. $\frac{3}{4} = \frac{\square}{8}$

c. $\frac{4}{5} = \frac{\square}{10}$

2. Draw horizontal lines to split each part of each fraction stick into 3 equal parts. Then fill in the missing numbers.

a. $\frac{1}{3} = \frac{\square}{9}$

b. $\frac{3}{4} = \frac{\square}{12}$

c. $\frac{4}{5} = \frac{12}{\square}$

3. Draw horizontal lines to split each part of each fraction stick into 4 equal parts. Then fill in the missing numbers.

a. 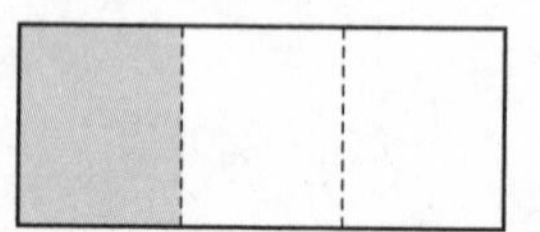

$\frac{1}{3} = \frac{\square}{12}$

b. 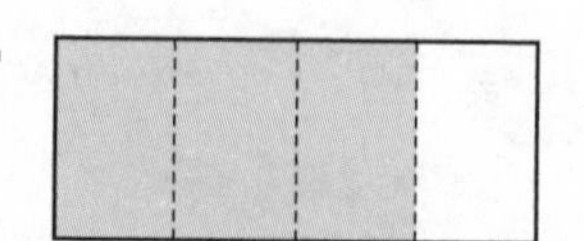

$\frac{3}{4} = \frac{12}{\square}$

c. 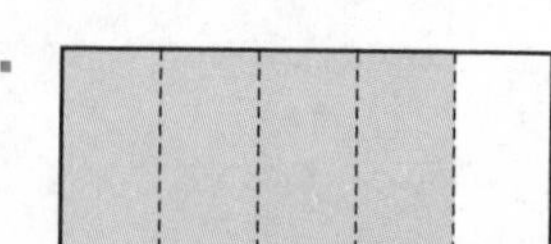

$\frac{4}{5} = \frac{\square}{\square}$

Date Time

Equivalent Fractions

Study the example below. Then solve Problems 1–3 in the same way. Match each fraction in the left column with an equivalent fraction in the right column.

Then fill in each box in the left column with a multiplication or division symbol and a number to show how each fraction is changed to get the equivalent fraction.

Example

$\frac{3 \ [*2]}{7 \ [*2]}$	$\frac{3}{6}$
$\frac{6 \ [\div 2]}{12 \ [\div 2]}$	$\frac{20}{30}$
$\frac{3 \ [\div 3]}{18 \ [\div 3]}$	$\frac{1}{6}$
$\frac{4 \ [*5]}{6 \ [*5]}$	$\frac{6}{14}$

1.

$\frac{1 \ [\ ]}{6 \ [\ ]}$	$\frac{6}{24}$
$\frac{6 \ [\ ]}{9 \ [\ ]}$	$\frac{2}{12}$
$\frac{1 \ [\ ]}{4 \ [\ ]}$	$\frac{3}{8}$
$\frac{6 \ [\ ]}{16 \ [\ ]}$	$\frac{2}{3}$

2.

$\frac{1 \ [\ ]}{2 \ [\ ]}$	$\frac{25}{30}$
$\frac{16 \ [\ ]}{24 \ [\ ]}$	$\frac{1}{3}$
$\frac{5 \ [\ ]}{6 \ [\ ]}$	$\frac{6}{12}$
$\frac{10 \ [\ ]}{30 \ [\ ]}$	$\frac{4}{6}$

3.

$\frac{15 \ [\ ]}{20 \ [\ ]}$	$\frac{28}{30}$
$\frac{6 \ [\ ]}{10 \ [\ ]}$	$\frac{4}{5}$
$\frac{12 \ [\ ]}{15 \ [\ ]}$	$\frac{3}{4}$
$\frac{14 \ [\ ]}{15 \ [\ ]}$	$\frac{12}{20}$

Date Time

Math Boxes 5.4

1. Find the landmarks for this set of numbers: 99, 87, 85, 32, 57, 82, 85, 99, 85, 65, 78, 87, 85, 57, 85, 99

   a. Maximum: ______

   b. Minimum: ______

   c. Range: ______

   d. Median: ______

2. Solve mentally.

   a. $299 * 50 =$ ______

   b. $1,999 * 4 =$ ______

   c. $99 * 72 =$ ______

   d. ______ $= 80 * 29$

   e. ______ $= 49 * 60$

3. Write a number story for the number model $743 / 8 = n$. Then solve it.

______

______

______

______

Answer: ______

4. Draw two different rectangles on the grid below, each with a perimeter of 16 units.

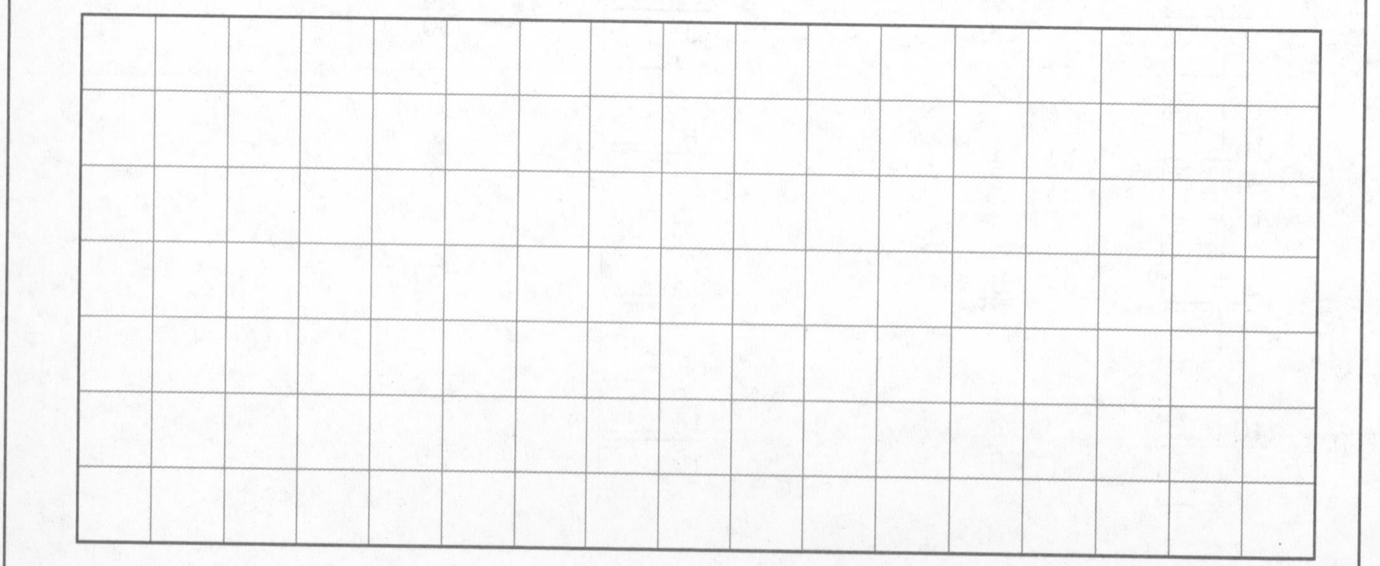

Date Time

Renaming Fractions as Decimals

1. Fill in the missing numbers and shade the squares. Each large square is worth 1.

Whole
large square

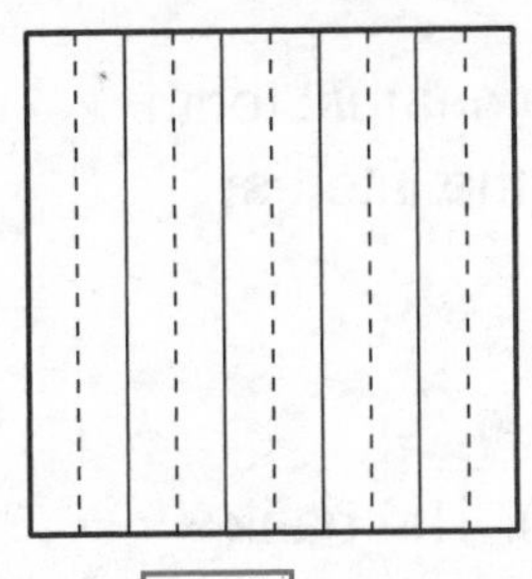

$\frac{2}{5} = \frac{\square}{10} = 0.$____

Shade $\frac{2}{5}$ of the square.

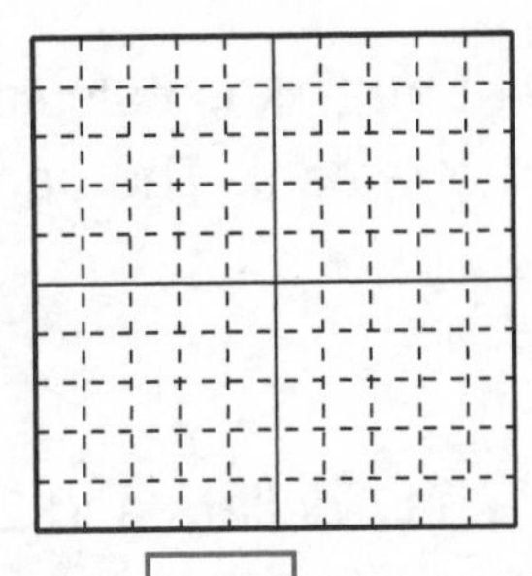

$\frac{3}{4} = \frac{\square}{100} = 0.$____

Shade $\frac{3}{4}$ of the square.

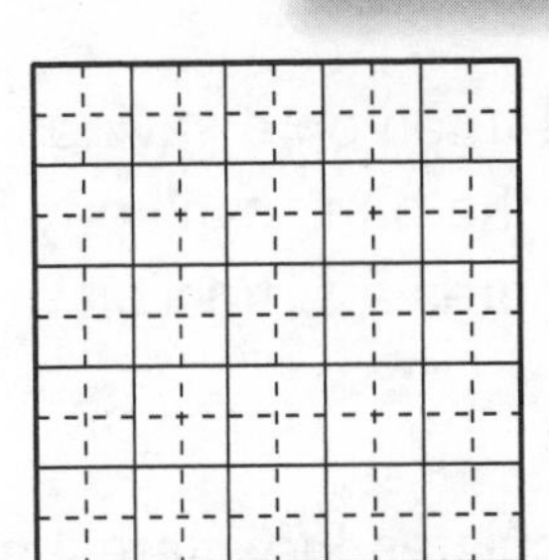

$\frac{15}{25} = \frac{\square}{100} = 0.$____

Shade $\frac{15}{25}$ of the square.

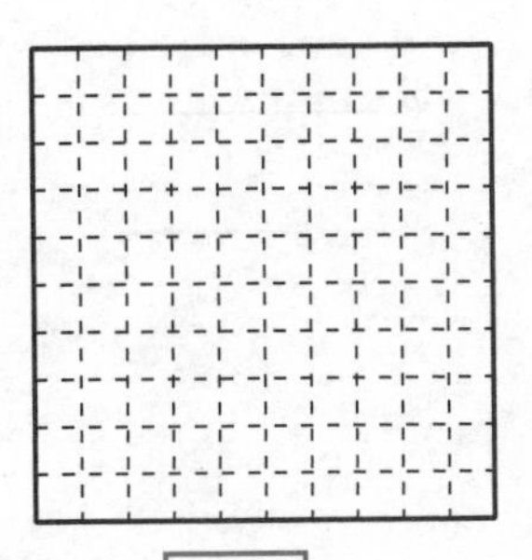

$\frac{3}{20} = \frac{\square}{100} = 0.$____

Shade $\frac{3}{20}$ of the square.

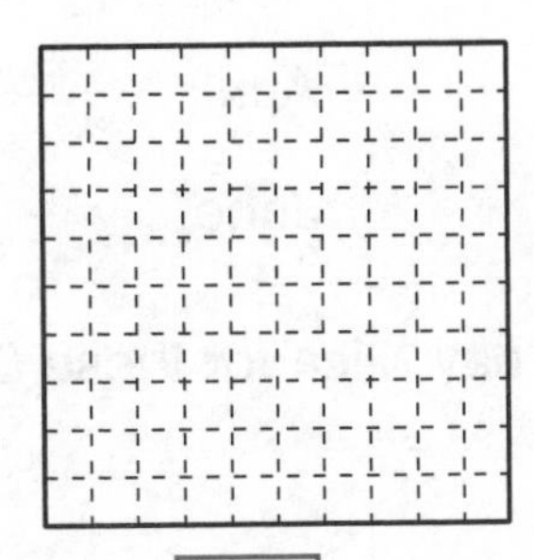

$\frac{8}{50} = \frac{\square}{100} = 0.$____

Shade $\frac{8}{50}$ of the square.

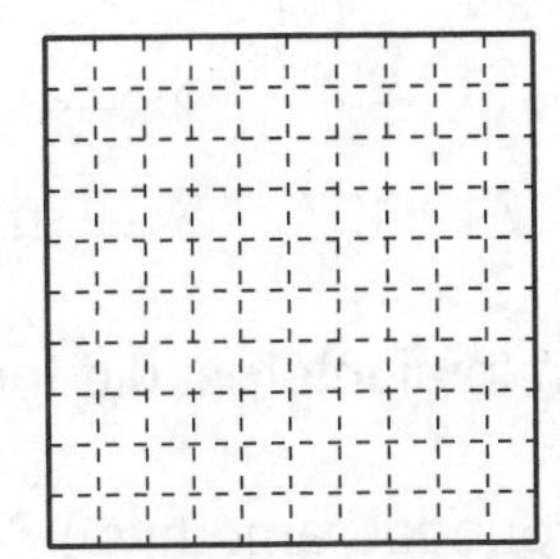

Write the shaded part as a fraction and as a decimal.

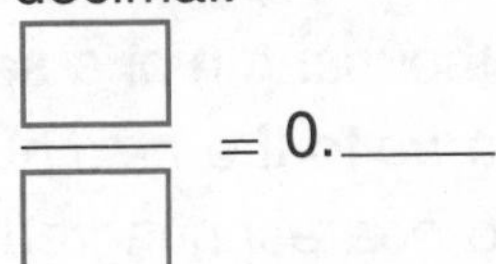

$\frac{\square}{\square} = 0.$____

2. Write each number below as a decimal. Then use the letters to locate the decimals on the number line.

a. $\frac{1}{2}$ = 0.5 **b.** $\frac{6}{10}$ = . **c.** $\frac{4}{5}$ = . **d.** $\frac{23}{100}$ = .

e. $\frac{22}{25}$ = . **f.** $\frac{21}{50}$ = . **g.** $\frac{7}{5}$ = . **h.** $1\frac{15}{50}$ = .

a

0 0.1 0.2 0.3 0.4 0.5 0.6 0.7 0.8 0.9 **1.0** 1.1 1.2 1.3 1.4 1.5

Date Time

Rounding Decimals

Sometimes numbers have more digits than are needed. Many calculators give answers to eight or more decimal places, even though only one or two places make sense. Rounding is a way to get rid of extra digits.

The interest earned on a savings account at a bank is calculated to the nearest tenth of a cent. But the bank can't pay a fraction of a cent. The bank *rounds* the interest *down,* and ignores any fraction of a cent.

Example

The bank calculates the interest as $17.218 (17 dollars and 21.8 cents). The bank ignores the 0.8 (or $\frac{8}{10}$) cent. It pays $17.21 in interest.

1. The calculated interest on Mica's savings account for 6 months is listed below. Round each amount down to find the interest actually paid each month.

January	$21.403	$ ________	February	$22.403	$ ________
March	$18.259	$ ________	April	$19.024	$ ________
May	$17.427	$ ________	June	$18.916	$ ________

How much total interest did the bank pay Mica for these 6 months?

(Add the rounded amounts.) $ ________

At the Olympic Games, each running event is timed to the nearest thousandth of a second. The timer *rounds* the time *up* to the *next* hundredth of a second (not the *nearest* hundredth). The rounded time becomes the official time.

Examples

11.437 seconds is rounded up to 11.44 seconds.

11.431 seconds is rounded up to 11.44 seconds.

11.430 seconds is reported as 11.43 seconds, since 11.430 is equal to 11.43.

Michael Johnson with his record-breaking time

Date Time

Rounding Decimals (cont.)

2. Find the official times for these runs. s: second(s) min: minute(s)

Electric Timer	Official Time	Electric Timer	Official Time
10.752 s	___.___ ___ s	20.001 s	___.___ ___ s
11.191 s	___.___ ___ s	43.505 s	___.___ ___ s
10.815 s	___.___ ___ s	49.993 s	___.___ ___ s
21.970 s	___.___ ___ s	1 min 55.738 s	___ min ___.___ s
20.092 s	___.___ ___ s	1 min 59.991 s	___ min ___.___ s

3. Describe a situation involving money when the result of a computation might always be rounded up.

____

____

____

Supermarkets often show unit prices for items. This helps customers comparison shop. A unit price is found by dividing the price of an item (in cents, or dollars and cents) by the quantity of the item (often in pounds). When the quotient has more decimal places than are needed, it is *rounded to the nearest* tenth of a cent.

Examples

23.822 cents (per ounce) is rounded down to 23.8 cents.

24.769 cents is rounded up to 24.8 cents.

18.65 cents is halfway between 18.6 cents and 18.7 cents. It is rounded up to 18.7 cents.

4. Round these unit prices to the nearest tenth of a cent (per ounce).

a. 28.374¢ ________ ¢ **b.** 19.796¢ ________ ¢ **c.** 29.327¢ ________ ¢

d. 16.916¢ ________ ¢ **e.** 20.641¢ ________ ¢ **f.** 25.583¢ ________ ¢

g. 18.469¢ ________ ¢ **h.** 24.944¢ ________ ¢ **i.** 17.281¢ ________ ¢

j. 23.836¢ ________ ¢ **k.** 21.866¢ ________ ¢ **l.** 22.814¢ ________ ¢

Date Time

Math Boxes 5.5

1. The trapezoid on your Geometry Template is worth 1. Use your template to draw a shape worth $2\frac{1}{3}$.

2. Complete the "What's My Rule?" table and state the rule.

Rule:

____

in	out
27	20
	6
5	−2
10	

3. Fran had $6.48 to spend on lunch. She bought a hamburger for $2.83. How much did she have left to spend after buying the hamburger?

____

4. Put the following fractions in order from least to greatest.

$\frac{3}{8}$ $\frac{4}{5}$ $\frac{2}{3}$ $\frac{1}{4}$ $\frac{9}{10}$

_____ , _____ , _____ , _____ , _____

5. Subtract.

a. $\begin{array}{r} 215 \\ -\ 38 \\ \hline \end{array}$

b. $\begin{array}{r} 309 \\ -\ 87 \\ \hline \end{array}$

c. $\begin{array}{r} 454 \\ -\ 376 \\ \hline \end{array}$

d. $\begin{array}{r} 270 \\ -\ 56 \\ \hline \end{array}$

Date Time

Writing Fractions as Decimals

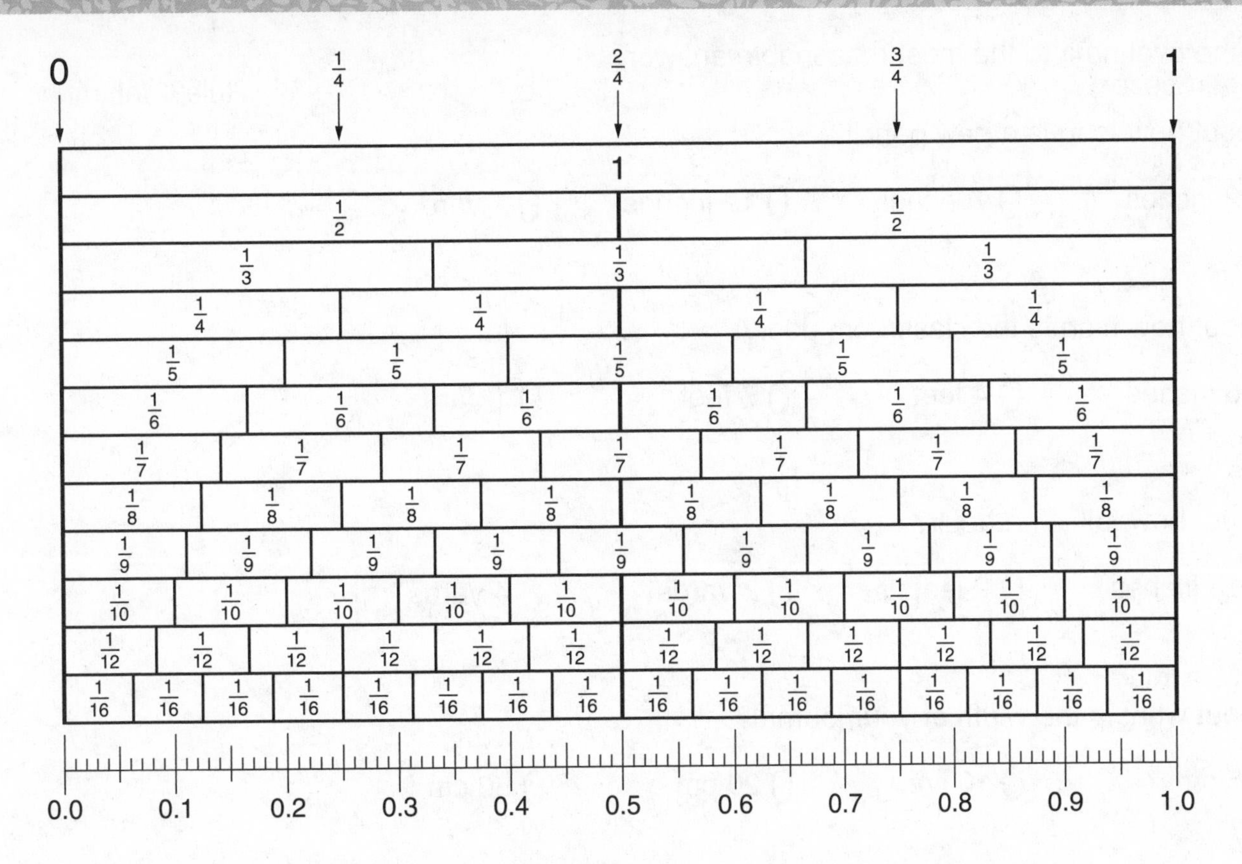

Use a straightedge and the above chart to fill in the blanks to the right of each fraction below. Write a decimal that is equal to, or about equal to, the given fraction. Directions for filling in the blank to the left of each fraction will be given in the next lesson.

_____ $\frac{1}{3}$ = 0. 3 3

_____ $\frac{2}{3}$ = 0.___ ___

_____ $\frac{4}{10}$ = 0.___ ___

_____ $\frac{4}{5}$ = 0.___ ___

_____ $\frac{1}{8}$ = 0.___ ___

_____ $\frac{5}{8}$ = 0.___ ___

_____ $\frac{9}{12}$ = 0.___ ___

_____ $\frac{11}{12}$ = 0.___ ___

_____ $1\frac{1}{3}$ = 1. 3 3

_____ $1\frac{3}{8}$ = 1.___ ___

_____ $3\frac{7}{8}$ = 3.___ ___

_____ $9\frac{5}{6}$ = ___.___ ___

Date Time

Measurement Review

Fill in the oval next to the most reasonable answer.

1. About how long is a new pencil?

O 2 inches O 7 inches O 12 inches O 1 yard

2. About how high is the classroom door?

O 6 inches O 4 feet O 7 feet O 1 yard

3. About how tall is an adult?

O 18 inches O 2 feet O 2 yards O 4 yards

4. About what is the width of your journal?

O 5 cm O 10 cm O 20 cm O 50 cm

Fill in the oval next to the best unit to use for each measurement.

5. The weight of an ant

O ounce O kilogram O foot O cup

6. The amount of juice a 5-year-old drinks each day

O ounce O kilogram O foot O cup

7. The length of a boat

O ounce O kilogram O foot O cup

8. The weight of an elephant

O ounce O kilogram O foot O cup

Date Time

Measurement Review (cont.)

Measure each line segment to the nearest $\frac{1}{8}$-inch.

9. ________________________________

________________ inches

10. ______________________

________________ inches

Measure each line segment to the nearest $\frac{1}{16}$-inch.

11. ______________

________________ inches

12. ______________________________

________________ inches

Draw a line segment

13. 8 centimeters long.

14. 4.7 centimeters long.

Date Time

Math Boxes 5.6

1. a. Make up a set of at least twelve numbers that have the following landmarks.

Minimum: 50
Maximum: 57
Median: 54
Mode: 56

____

____

____

b. Make a bar graph for this set of numbers.

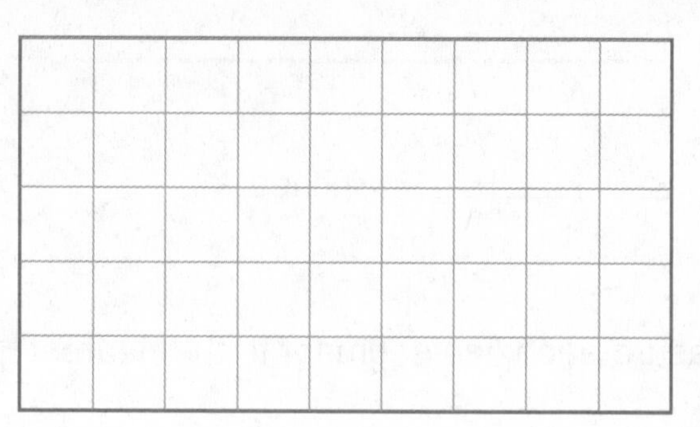

2. Complete the table.

Fraction	Decimal	Percent
$\frac{1}{3}$		
		30%
	0.65	
		40%
$\frac{1}{20}$		

3. Measure the length and width of each of the following objects to the nearest centimeter.

a. *Student Reference Book* cover

length: ____ cm width: ____ cm

b. seat of chair

length: ____ cm width: ____ cm

c. sole of shoe

length: ____ cm width: ____ cm

4. I am a number. If you double $\frac{1}{4}$ of me, you get 16. What number am I?

____

5. Write five names for 7.5.

____

____

____

____

____

Date Time

More about Writing Fractions as Decimals

How to Write a Repeating Decimal

Some decimal numbers use up the entire calculator display. If a digit repeats, the decimal number can be written in a simple way by putting a bar over the repeating digit. Study these examples.

Fraction	Divide Numerator by Denominator. Calculator Display:	Write the Decimal this Way:
$\frac{1}{3}$	0.3333333333	$0.\overline{3}$
$\frac{2}{3}$	0.6666666666 or 0.6666666667 (depending on the calculator)	$0.\overline{6}$
$\frac{1}{12}$	0.0833333333	$0.08\overline{3}$
$\frac{8}{9}$	0.8888888888 or 0.8888888889 (depending on the calculator)	$0.\overline{8}$

Use your calculator to convert each fraction below to a decimal by dividing. If the result is a repeating decimal, write a bar over the digit or digits that repeat. Then circle the correct answer to each question.

1. Which is closer to 0.8? $\frac{6}{8}$ ______________ or $\frac{5}{6}$ ______________
2. Which is closer to 0.25? $\frac{2}{9}$ ______________ or $\frac{3}{9}$ ______________
3. Which is closer to 0.6? $\frac{4}{7}$ ______________ or $\frac{7}{12}$ ______________
4. Which is closer to 0.05? $\frac{1}{30}$ ______________ or $\frac{1}{12}$ ______________
5. Which is closer to 0.39? $\frac{3}{8}$ ______________ or $\frac{7}{16}$ ______________

Date Time

Math Boxes 5.7

1. The large rhombus on your Geometry Template is worth 1. Use your Template to draw a shape worth $2\frac{1}{2}$.

2. Complete the "What's My Rule?" table and state the rule.

Rule:

____

in	out
8	17
11	
5	14
	4

3. Sophie went to the ball game. She spent $8.50 on the ticket, $2.75 on a hot dog, $1.99 on a soft drink, and $0.15 on a souvenir pencil. How much did she spend in all?

____

4. Put the following fractions in order from least to greatest.

$\frac{3}{7}$ $\frac{3}{5}$ $\frac{2}{8}$ $\frac{8}{9}$ $\frac{5}{6}$

_____ , _____ , _____ , _____ , _____

5. Subtract.

a. $\begin{array}{r} 727 \\ -\ 47 \\ \hline \end{array}$

b. $\begin{array}{r} 503 \\ -\ 65 \\ \hline \end{array}$

c. $\begin{array}{r} 248 \\ -\ 176 \\ \hline \end{array}$

d. $\begin{array}{r} 2,403 \\ -\ 764 \\ \hline \end{array}$

Date Time

Converting Fractions to Decimals and Percents

Example Teneil used her calculator to rename the following fraction as a decimal and as a percent.

$\frac{14}{23}$	14 ÷ 23 = 0.6086956522	100 × 0.6086956522 = 60.86956522%
Fraction	**Decimal**	**Percent**

Teneil only needed to work with a whole percent, so she rounded 60.86956522% to 61%.

1. Use your calculator to convert each fraction to a decimal. Write all of the digits shown in the display. Then write the equivalent percent rounded to the nearest whole percent. The first row has been done for you.

Fraction	Decimal	Percent (rounded to the nearest whole percent)
$\frac{18}{35}$	0.5142857143	51%
$\frac{12}{67}$		
$\frac{24}{93}$		
$\frac{13}{24}$		
$\frac{576}{1,339}$		

2. Linell got 80% correct on a spelling test. If the test had 20 questions, how many did Linell get correct? ________ questions

3. Jamie spent 50% of his money on a baseball cap. The cap cost $15. How much money did Jamie have at the beginning? ________

4. Hunter got 75% correct on a music test. If he got 15 questions correct, how many questions were on the test? ________ questions

Date Time

Converting Fractions to Decimals and Percents (cont.)

5. Below is a list of 10 animals and the average number of hours per day that each spends sleeping.

Write the fraction of a day that each animal sleeps. Then calculate the equivalent decimal and percent (rounded to the nearest whole percent). You may use your calculator. The first row has been done for you.

Animal	Average Hours of Sleep per Day	Fraction of Day Spent Sleeping	Decimal Equivalent	Percent of Day Spent Sleeping (to the nearest whole percent)
koala	22	$\frac{22}{24}$	$0.91\overline{6}$	92%
sloth	20			
armadillo and opossum	19			
lemur	16			
hamster and squirrel	14			
cat and pig	13			
spiny anteater	12			

Source: The Top 10 of Everything 2000

6. The total number of horses in the world is about 60,800,000. China is the country with the greatest number of horses (about 8,900,000). What percent of the world's horses live in China? ______________

7. In the United States, about 45% of the population has blood type O. About how many people out of every 100 have blood type O? ______________

8. About 11 out of every 100 households in the United States has a parakeet. How would you express this as a percent? ______________

Math Boxes 5.8

1. a. Make up a set of at least twelve numbers that have the following landmarks.

Minimum: 3
Maximum: 9
Median: 7
Mode: 7

____

____

____

b. Make a bar graph for this set of numbers.

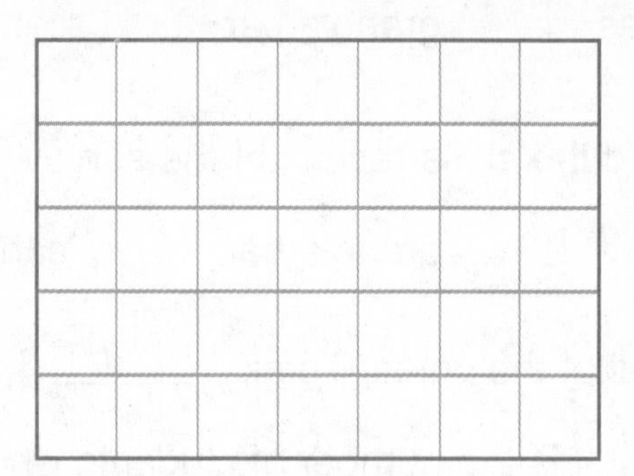

2. Complete the table.

Fraction	Decimal	Percent
		95%
	0.80	
$\frac{3}{9}$		
$\frac{6}{8}$		
		$66\frac{2}{3}\%$

3. Measure the length and width of each of the following objects to the nearest centimeter.

a. pinkie finger

length: ____ cm width: ____ cm

b. notebook

length: ____ cm width: ____ cm

c. pencil

length: ____ cm width: ____ cm

4. I am a number. If you double $\frac{1}{3}$ of me, you get 14. What number am I?

____

5. Write five names for 100.

____

____

____

____

____

Date Time

Bar Graphs and Circle (Pie) Graphs

1. Circle the after-school snack you like best. Mark only one answer.

cookies granola bar candy bar fruit other

2. Record the class results of the survey.

cookies ____ granola bar ____ candy bar ____ fruit ____ other ____

Add all of the votes. Total: ________

The total is the number of students who voted.

3. Make a bar graph showing the results.

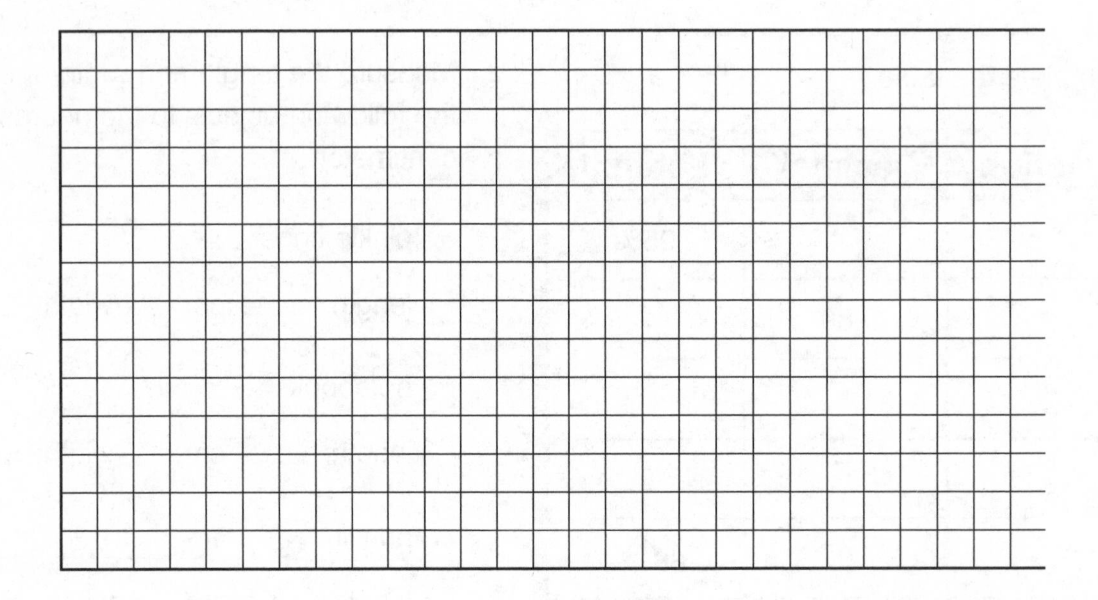

4. Another fifth grade class with 20 students collected snack-survey data. The class made the circle graph (also called a pie graph) below.

Tell how you think they made the graph.

____

____

____

____

____

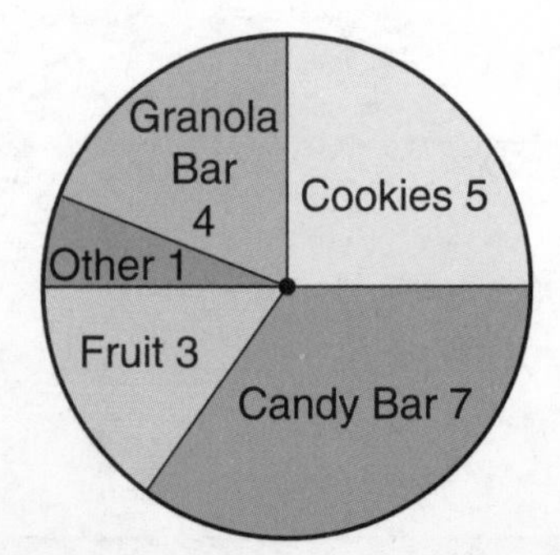

Date Time

Multiplication Practice

Use a favorite strategy (not a calculator) to multiply.

1. $\begin{array}{r} 48 \\ * \ 29 \\ \hline \end{array}$

2. $\begin{array}{r} 34 \\ * \ 79 \\ \hline \end{array}$

3. $\begin{array}{r} 62 \\ * \ 53 \\ \hline \end{array}$

4. $\begin{array}{r} 119 \\ * \ 47 \\ \hline \end{array}$

5. $\begin{array}{r} 305 \\ * \ 29 \\ \hline \end{array}$

6. $\begin{array}{r} 245 \\ * \ 51 \\ \hline \end{array}$

7. Reggie multiplied 28 * 73 with a lattice as shown below. Correct his mistakes and record the correct answer below.

	2	8	
1	1/4 (carry 1)	5/6 (carry 1)	7
5	6/6	2/0	3
	8	0	

28 * 73 = ______________

Date Time

Math Boxes 5.9

1. Circle all the fractions that are equivalent to $\frac{9}{18}$.

$\frac{7}{14}$ $\frac{7}{8}$ $\frac{6}{9}$ $\frac{5}{10}$ $\frac{2}{3}$

2. Write the prime factorization for each number.

a. 38 = ______________

b. 92 = ______________

c. 56 = ______________

d. 72 = ______________

e. 125 = ______________

3. Fill in the missing values on the number lines.

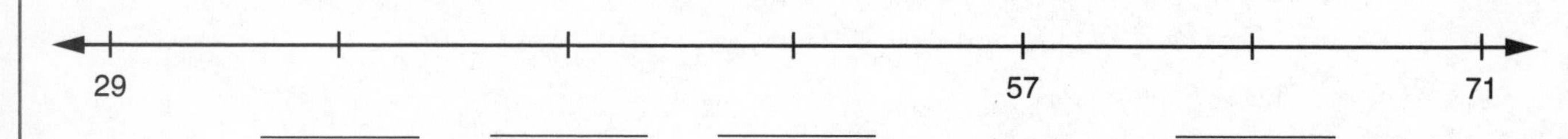

_______ _______ _______ _______

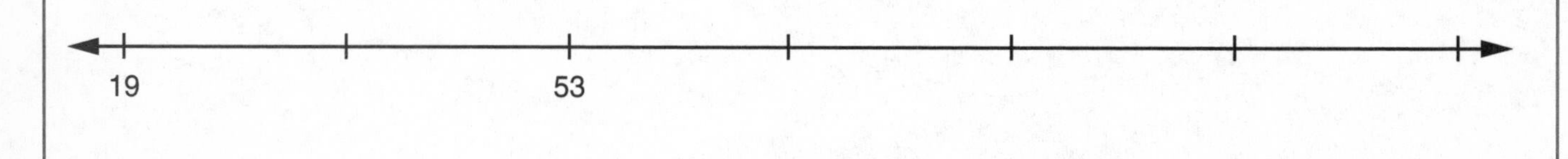

_______ _______ _______ _______ _______

4. Draw a circle with a radius of 3 centimeters.

What is the diameter of the circle?

______________ (unit)

SRB 143 152

Date Time

Reading Circle Graphs

Use your Percent Circle to find what percent each pie piece is of the whole circle.

1.

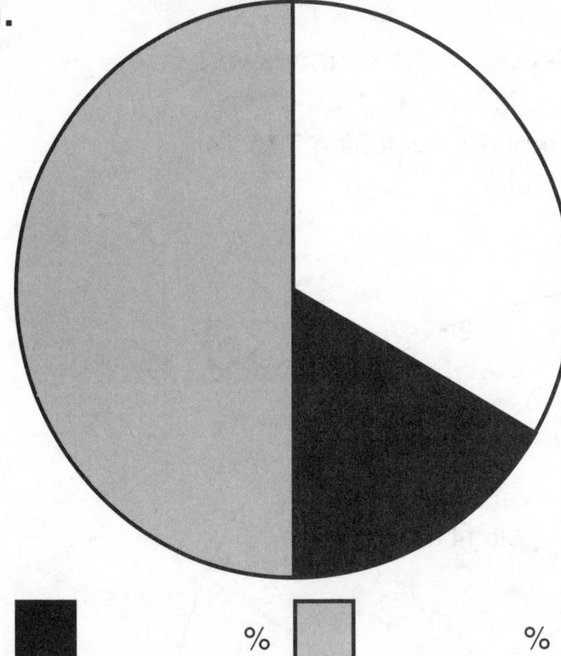

________ % ________ %

________ %

2.

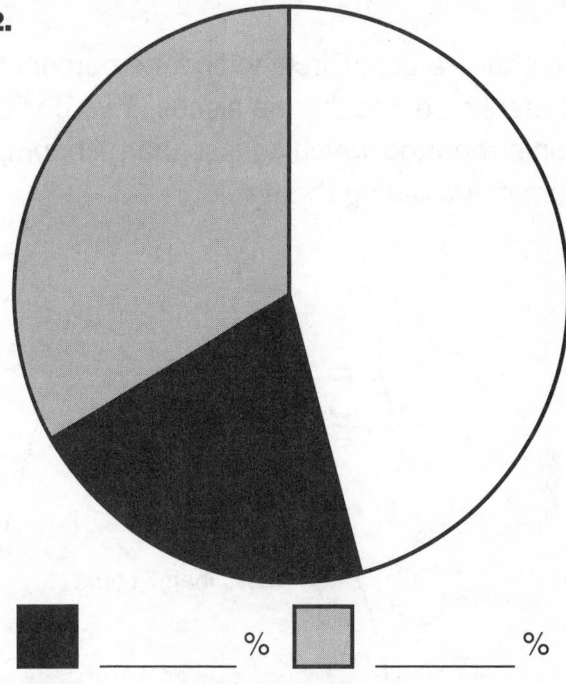

________ % ________ %

________ %

3.

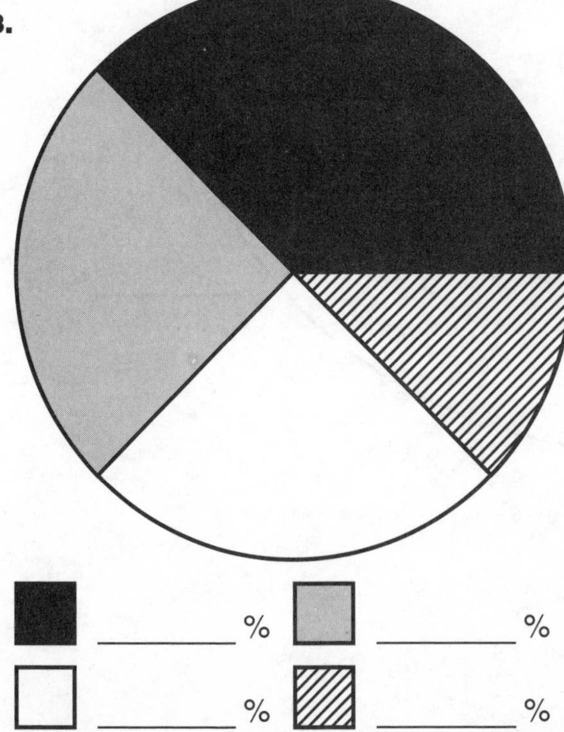

________ % ________ %

________ % ________ %

4.

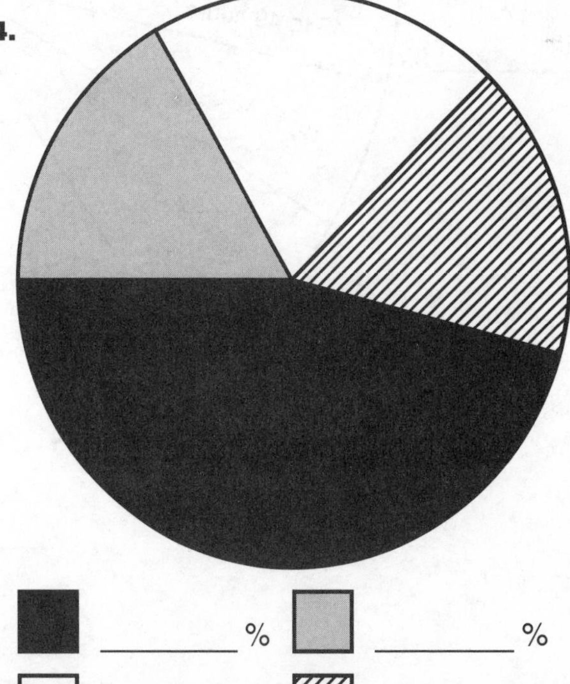

________ % ________ %

________ % ________ %

Date Time

How Much TV Do People Watch?

A large sample of people was asked to report on how much TV they watched during one week. The circle graph below shows the survey's results.

Use your Percent Circle to find the percent in each category. Write your answers in the blanks next to the pie pieces. Two percents are filled in for you: 18% of the people reported watching less than 7 hours; and 30% reported watching 7 to 14 hours of TV during the week.

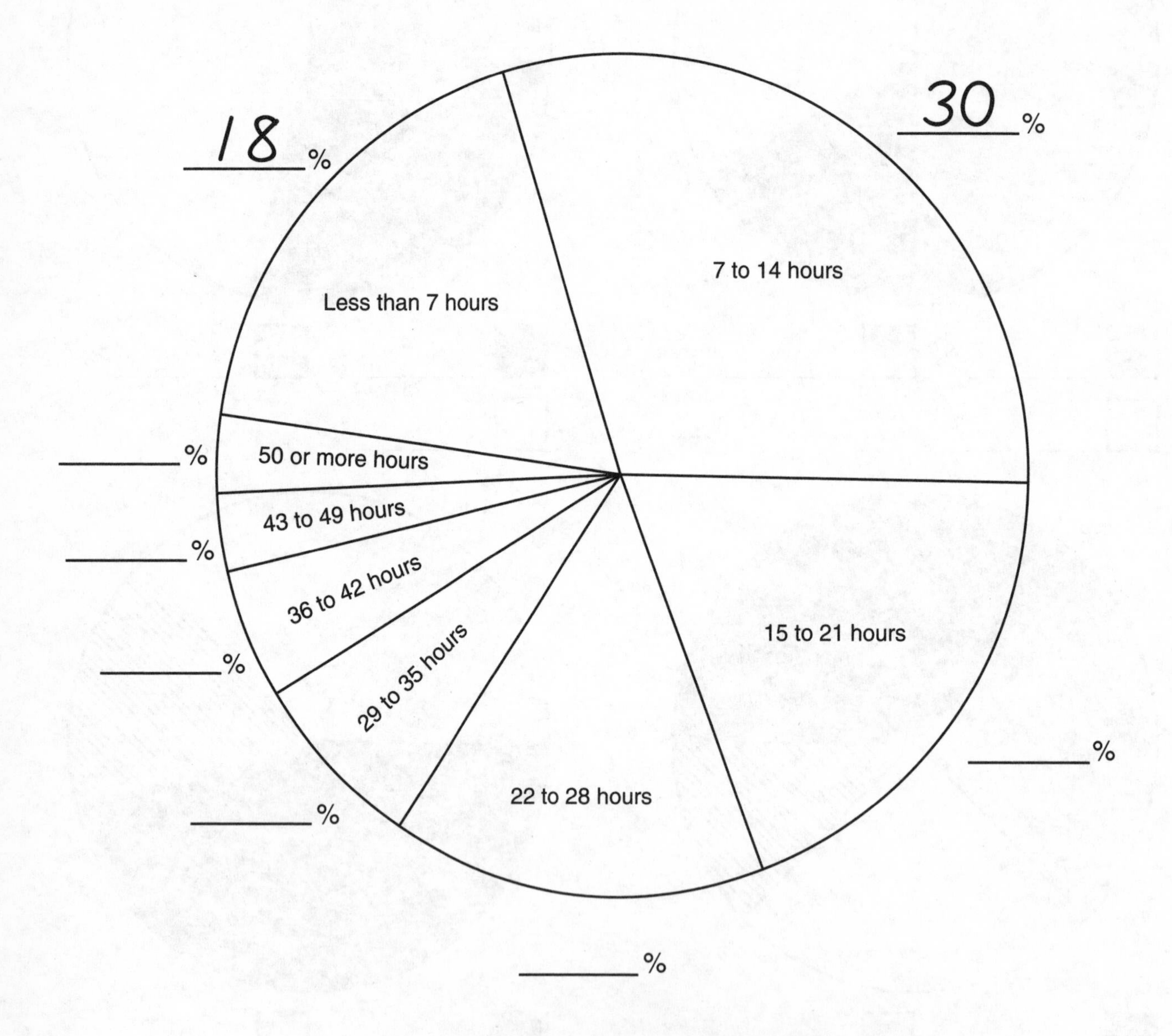

_____ %

Date Time

Division Practice

Estimate each quotient. Solve only the problems with a quotient that is less than 200. Use a favorite strategy (not a calculator) to divide.

1. $5\overline{)684}$ ____________ **2.** $7\overline{)329}$ ____________ **3.** $4\overline{)994}$ ____________

4. $6\overline{)637}$ ____________ **5.** $9\overline{)1{,}243}$ ____________ **6.** $5\overline{)1{,}585}$ ____________

Date Time

Math Boxes 5.10

1. Complete.

   a. 1 hour = ______________ minutes

   b. 3 hours = ______________ minutes

   c. 5 weeks = ______________ days

   d. 4 years = ______________ months

   e. $2\frac{1}{2}$ years = ______________ months

2. Round each number to the nearest hundredth.

   a. 3.130 ______________

   b. 10.647 ______________

   c. 29.999 ______________

   d. 45.056 ______________

   e. 87.708 ______________

3. Write > or <.

   a. $\frac{1}{4}$ ________ $\frac{3}{8}$

   b. $\frac{2}{7}$ ________ $\frac{2}{5}$

   c. $\frac{8}{9}$ ________ $\frac{7}{8}$

   d. $\frac{7}{12}$ ________ $\frac{3}{6}$

   e. $\frac{5}{12}$ ________ $\frac{5}{11}$

4. Add or subtract. Show your work.

   a. 2.03 − 0.76 = ______________

   b. ______________ = 57.97 + 3.03

   c. ______________ = 691.23 − 507.26

   d. ______________ = 29.05 + 103.94

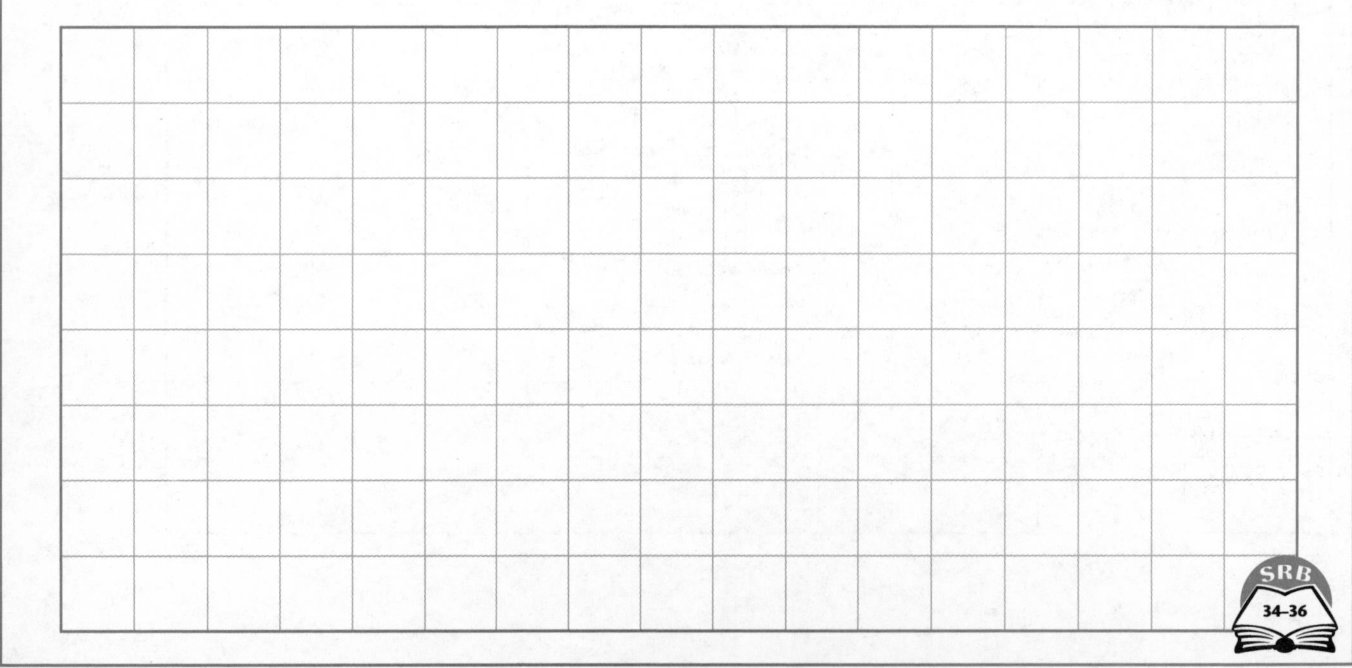

Date Time

Making Circle Graphs: Concrete

Concrete is an artificial stone. It is made by first mixing cement and sand with gravel or other broken stone. Then enough water is mixed in to cause the cement to set. After drying (or curing), the result is a hard piece of concrete.

The cement, sand, and gravel are commonly mixed using this recipe:

Recipe for Dry Mix for Concrete		
Material	**Fractional Part of Mix**	**Percent Part of Mix**
Cement	$\frac{1}{6}$	$16\frac{2}{3}\%$
Sand	$\frac{1}{3}$	$33\frac{1}{3}\%$
Gravel	$\frac{1}{2}$	50%

Use your Percent Circle to make a circle graph for the above recipe in the circle below. Label each section of the graph, and give it a title.

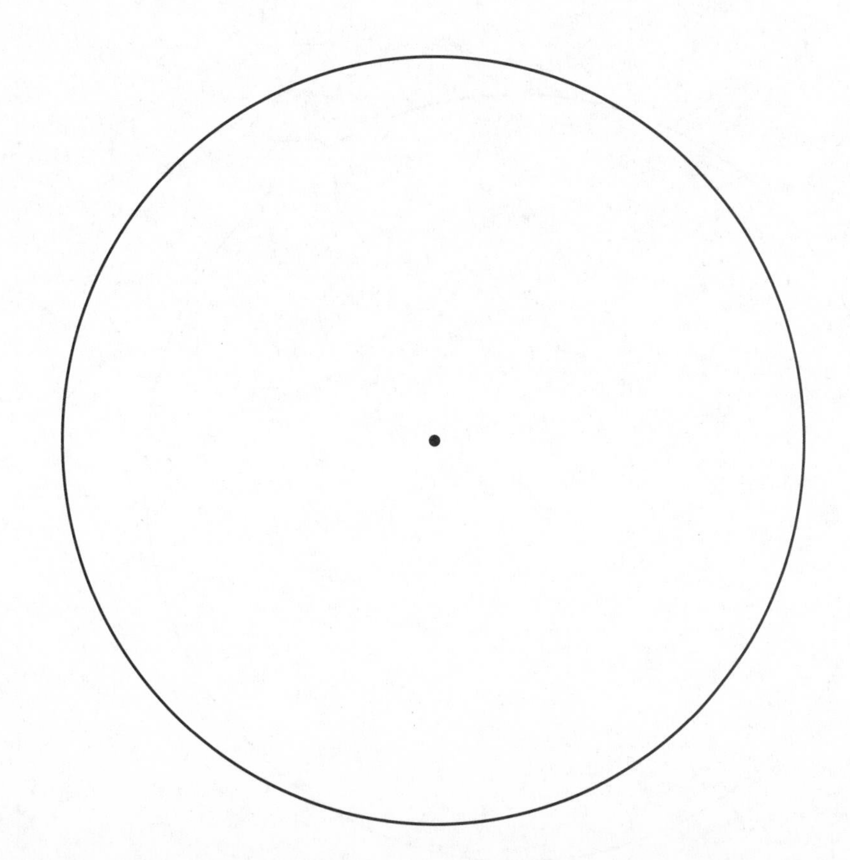

Date Time

Making Circle Graphs: Snack Survey

Your class recently made a survey of favorite snacks. As your teacher tells you the percent of votes each snack received, record the data in the table at the right. Make a circle graph of the snack-survey data in the circle below. Remember to label each piece of the graph and give it a title.

Votes			
Snack	**Number**	**Fraction**	**Percent**
Cookies			
Granola Bar			
Candy Bar			
Fruit			
Other			
Total			About 100%

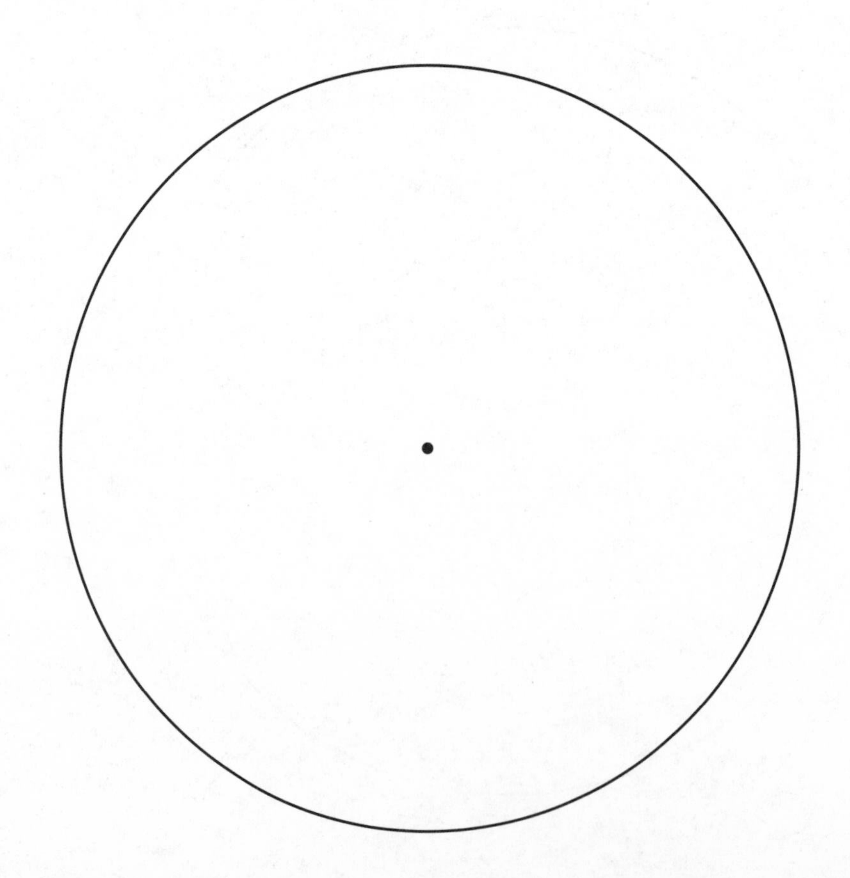

Date Time

Math Boxes 5.11

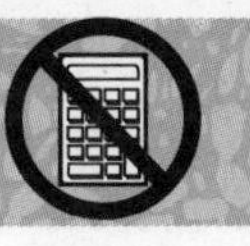

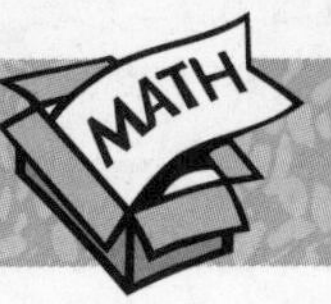

1. Circle all the fractions that are equivalent to $\frac{4}{12}$.

$\frac{5}{15}$ $\frac{2}{6}$ $\frac{8}{16}$ $\frac{3}{9}$ $\frac{12}{16}$

2. Write the prime factorization for each number.

a. 90 = ______

b. 54 = ______

c. 75 = ______

d. 112 = ______

e. 88 = ______

3. Fill in the missing values on the number lines.

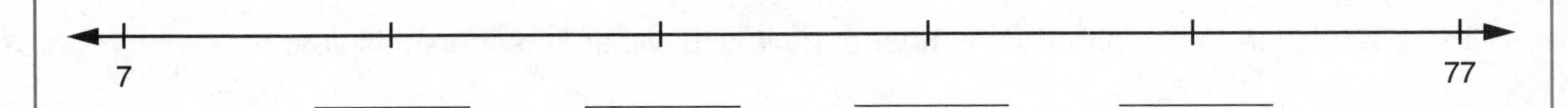

______ ______ ______ ______

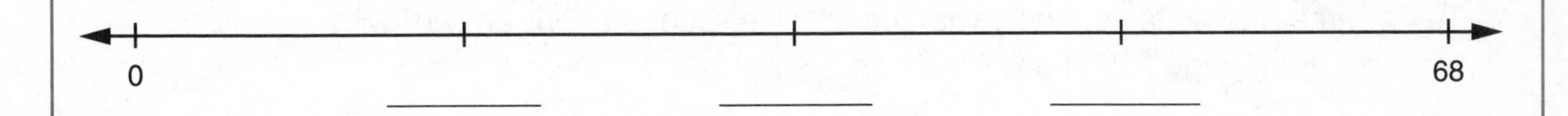

______ ______ ______

4. Draw a circle with a radius of 1 inch.

What is the diameter of the circle?

______
(unit)

Date Time

School Days

Read the article "School" on pages 318–320 in the American Tour section of the *Student Reference Book.*

1. Tell whether the statement below is true or false. Support your answer with evidence from page 318 of the American Tour.

   In 1790, it was common for 11-year-olds to go to school fewer than 90 days a year.

   ____

   ____

   ____

   ____

2. About how many days will you go to school this year? About ________ days

   Write a fraction to compare the number of days you will go to school this year to

   the number of days an 11-year-old might have gone to school in 1790. ________

3. Tell whether the statement below is true or false. Support your answer with evidence from page 319 of the American Tour.

   In 1900, students in some states spent twice as many days in school, on average, as students in some other states.

   ____

   ____

   ____

   ____

4. In 1900, in which region (Northeast, South, Midwest, or West) did students go to school …

   the greatest number of days per year? ________________

   the fewest number of days per year? ________________

 Use with Lesson 5.12.

Date Time

School Days (cont.)

Tell whether each statement below is true or false. Support your answer with evidence from the graphs on page 320 of the American Tour.

5. On average, students in 2000 were absent from school about one-third as many days as students were absent in 1900.

____

____

____

6. The average number of days students spent in school per year has not changed much since 1960.

____

____

____

Challenge

7. Tell whether the statement below is true or false. Support your answer with evidence from the American Tour.

   From 1900 to 1980, the average number of days students spent in school per year more than doubled.

____

____

____

8. Locate your state in the table "Average Number of Days in School per Student, 1900" on page 319 of the American Tour. If you are in Alaska or Hawaii, choose another state.

   Was your state above or below the median for its region? ______________

9. Locate the number of days in school for your state in the stem-and-leaf plot on page 319 of the American Tour.

   Was your state above or below the median for all states? ______________

Date Time

A Short History of Mathematics Instruction

Throughout our nation's history, students have learned mathematics in different ways and have spent their time working on different kinds of problems. This is because people's views of what students can and should learn are constantly changing.

1. **1790s** If you went to elementary school in 1790, you were probably not taught mathematics. People believed that it was too hard to teach mathematics to children younger than 12.

   Older students spent most of their time solving problems about buying and selling goods. Here is a typical problem for a student in high school or college in the 1700s. Try to solve it.

   If 7 yards of cloth cost 21 shillings (a unit of money), how much do 19 yards of cloth cost? ________ shillings

2. **1840s** It was discovered that children could be very good at mental arithmetic, and students began to solve mental arithmetic problems as early as age 4. A school in Connecticut reported that its arithmetic champion could mentally multiply 314,521,325 by 231,452,153 in $5\frac{1}{2}$ minutes.

   After studying arithmetic two hours a day for 7 to 9 years, 94% of eighth graders in Boston in 1845 could solve the following problem. Try to solve it.

   What is $\frac{1}{2}$ of $\frac{1}{3}$ of 9 hours, 18 minutes? ________________

3. **1870s** Many textbooks were step-by-step guides on how to solve various problems. Students were given problems and answers. They had to show how the rules in the textbook could be used to produce the given answers.

   Here is a problem from around 1870 (without the answer) given to students at the end of 6 to 8 years of elementary arithmetic study. Try to solve it.

   I was married at the age of 21. If I live 19 years longer, I will have been married 60 years. What is my age now? ________________ (units)

A Short History of Mathematics Instruction (cont.)

4. **1920s** Elementary mathematics emphasized skill with paper-and-pencil algorithms. People were needed to keep track of income, expenses, and profits for businesses. Clerks in stores had to add up sales, but there were no cheap, easy-to-use calculators. As a result, students spent much of their time doing exercises like the following. These problems are from a test for students in grades 5 through 8. Most students couldn't solve them until seventh grade. See how well you can do now (without a calculator).

$$\begin{array}{r} \$\ 0.49 \\ 0.28 \\ 0.63 \\ 0.95 \\ 1.69 \\ 0.22 \\ 0.33 \\ 0.36 \\ 1.01 \\ +\ 0.56 \\ \hline \end{array} \qquad \begin{array}{r} \$\ 8.00 \\ 5.75 \\ 2.33 \\ 4.16 \\ 0.94 \\ +\ 6.32 \\ \hline \end{array}$$

5. **1990s** Today the emphasis is on solving problems and applying mathematics in the everyday world. The following problem was solved correctly by 47% of eighth graders on a test given in 1990. Try to solve it.

The cost to rent a motorbike is given by the following formula:
Cost = ($3 * number of hours rented) + $2

Complete the following table:

Time	Cost
1 hour	$5
4 hours	$________
________ hours	$17

Date Time

Math Boxes 5.12

1. Complete.

a. $\frac{1}{2}$ hour = ______________ minutes

b. $\frac{2}{6}$ hour = ______________ minutes

c. $1\frac{1}{2}$ hours = ______________ minutes

d. $3\frac{1}{2}$ days = ______________ hours

e. 2 years = ______________ weeks

2. Round each number to the nearest tenth.

a. 18.19 = ______________

b. 50.243 = ______________

c. 79.999 = ______________

d. 62.081 = ______________

e. 25.008 = ______________

3. Write > or <.

a. $\frac{3}{8}$ ________ $\frac{3}{4}$

b. $\frac{9}{10}$ ________ $\frac{9}{16}$

c. $\frac{6}{7}$ ________ $\frac{5}{7}$

d. $\frac{10}{12}$ ________ $\frac{4}{6}$

e. $\frac{8}{9}$ ________ $\frac{6}{7}$

4. Add or subtract. Show your work.

a. 14.59 + 202.7 = ______________

b. 89 + 36.02 = ______________

c. ______________ = 60.07 − 0.08

d. ______________ = 15.76 − 5.99

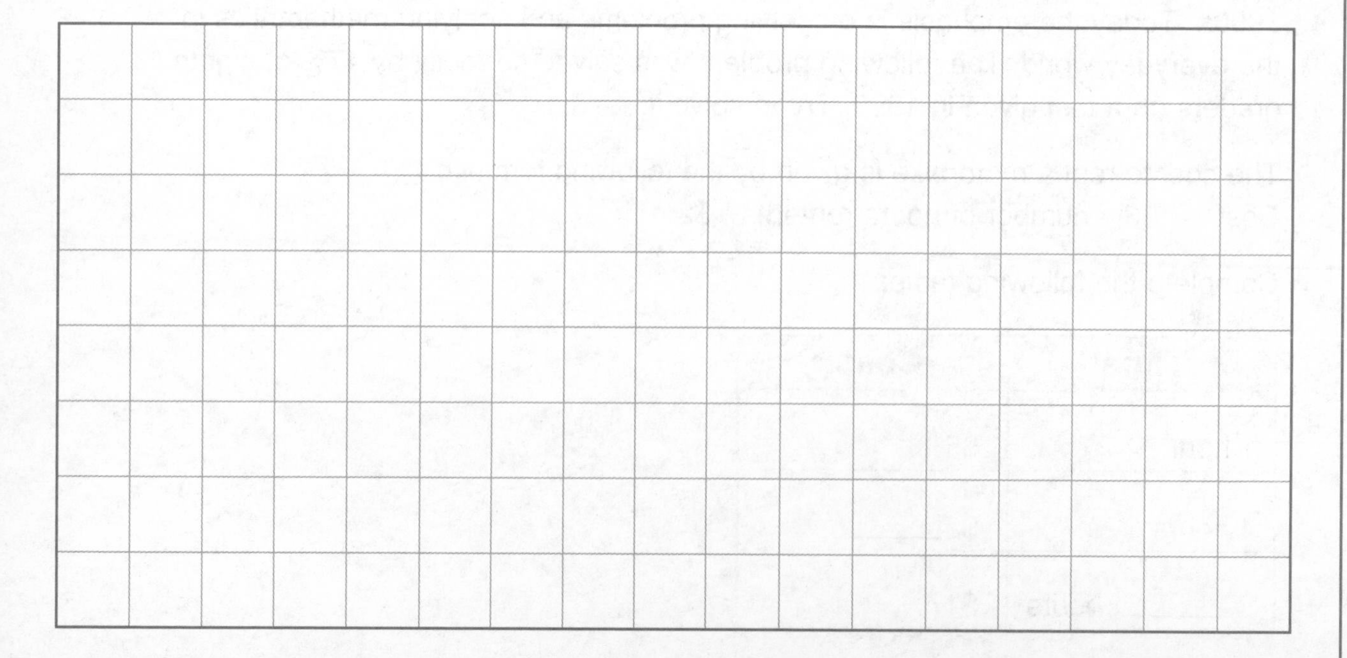

Date Time

Time to Reflect

1. Name two places outside of school where people use fractions.

____

____

In some situations, parts of a whole are usually named with fractions.
In some situations, they are usually named with decimals.
In some situations, they are usually named with percents.
Give at least one example of each type of situation below.

2. Parts of a whole usually named with a fraction:

____

____

3. Parts of a whole usually named with a decimal:

____

____

4. Parts of a whole usually named with a percent:

____

____

5. Explain one advantage to reporting test scores as a percent instead of as a fraction.

____

____

____

Date Time

Math Boxes 5.13

1. Complete.

a. $\frac{1}{4}$ hour = ______________ minutes

b. 20 minutes = ______________ hour

c. 30 minutes = ______________ hour

d. $\frac{3}{4}$ hour = ______________ minutes

e. $\frac{1}{12}$ hour = ______________ minutes

2. Express each of the following as a fraction, a mixed number, or a whole number.

= 2

= $1\frac{1}{2}$

= ________

= ________

= ________

3. a. Make up a set of at least twelve numbers that has the following landmarks.

Minimum: 28
Maximum: 34
Median: 30
Mode: 29

____

____

____

b. Make up a bar graph for this set of numbers.

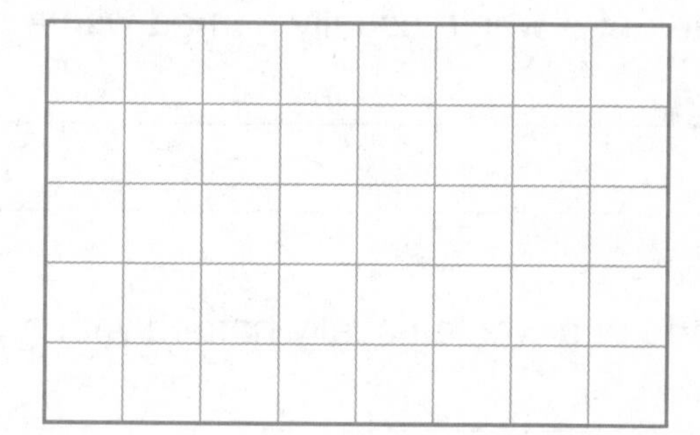

4. Complete the table.

Fraction	Decimal	Percent
$\frac{1}{5}$		
		38%
	0.75	
$\frac{4}{6}$		
		62.5%

5. Circle the fractions that are equivalent to $\frac{2}{3}$.

$\frac{10}{15}$ $\frac{4}{9}$ $\frac{9}{12}$ $\frac{12}{18}$ $\frac{4}{6}$

Date Time

Math Boxes 6.1

1. Write a 10-digit numeral that has
9 in the tens place,
3 in the millions place,
5 in the billions place,
7 in the hundred-millions place,
1 in the thousands place, and
6 in all other places.

_____ , _____ _____ _____ , _____ _____ _____ , _____ _____ _____

Write the numeral in words.

____

____

2. Round each number to the nearest hundredth.

a. 15.159 ____________

b. 8.003 ____________

c. 72.606 ____________

d. 964.443 ____________

e. 10.299 ____________

3. Write a fraction or a mixed number for each of the following:

a. 15 minutes = ________ hour

b. 40 minutes = ________ hour

c. 45 minutes = ________ hour

d. 25 minutes = ________ hour

e. 12 minutes = ________ hour

4. Rename each fraction as a mixed number or a whole number.

a. $\frac{28}{4}$ = ________

b. $\frac{36}{6}$ = ________

c. $\frac{25}{12}$ = ________

d. $\frac{46}{8}$ = ________

e. $\frac{18}{5}$ = ________

5. Complete.

a. ________ $*$ 600 = 24,000

b. ________ = 90 $*$ 90

c. ________ $*$ 20 = 1,000

d. ________ $*$ 70 = 49,000

e. 200,000 = 500 $*$ ________

Date Time

States Students Have Visited

1. You and your classmates counted the number of states each of you has visited. As the counts are reported and your teacher records them, write them in the space below. When you finish, circle your own count in the list.

2. Decide with your group how to organize the data you just listed. (For example, you might make a line plot or a tally table.) Then organize the data and show the results below.

3. Write two things you think are important about the data.

 a. ______________________________

 b. ______________________________

4. Compare your own count of states with those of your classmates.

____

____

Use with Lesson 6.1.

Date Time

States Adults Have Visited

1. You and your classmates each recorded the number of states that an adult had been in. As the numbers are reported and your teacher records them, write them in the space below.

2. Draw a line plot to organize the data you just listed.

3. Record landmarks for the data about adults and students in the table below.

Landmark	Adults	Students
Minimum		
Maximum		
Mode(s)		
Median		

4. How are the counts for adults and students different? Explain your answer.

Date Time

A Complicated Pizza

The pizza shown has been cut into 12 equal slices.

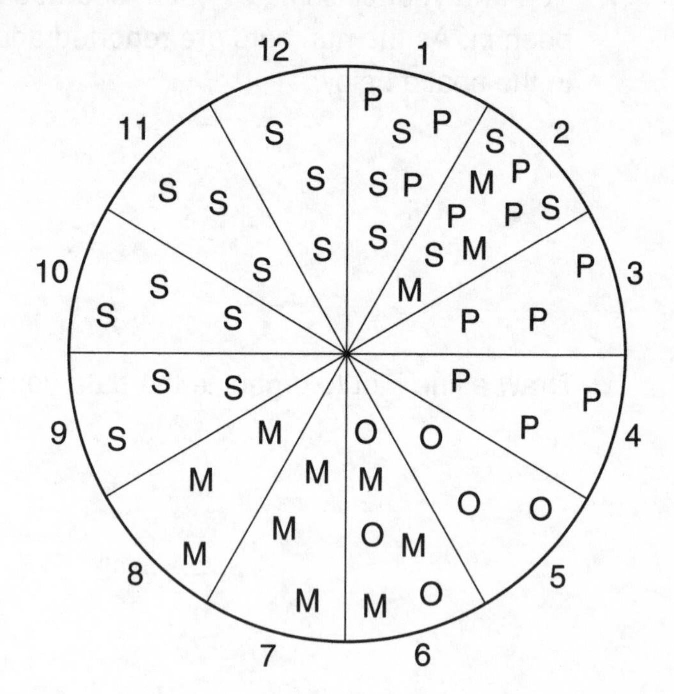

1. Fill in each blank with a fraction.
(*Hint:* Color-coding the pizza may help.)

______________ of the slices have ***just one*** type of topping.

______________ of the slices have ***2 or more*** types of toppings.

______________ of the slices have ***only*** sausage.

______________ of the slices have sausage as ***at least one*** topping.

______________ of the slices have ***no*** vegetables.

______________ of the slices have ***both*** meat and vegetables.

2. Suppose that all the slices with pepperoni are eaten first.

How many slices remain? ______________

What fraction of the slices remaining have mushrooms? ______________

What fraction of the slices remaining have only mushrooms? ______________

3. Bob, Sara, Don, and Alice share the pizza. Each person will eat exactly 3 slices.

Bob will eat slices with only meat (sausage and pepperoni). Alice will eat slices with only vegetables (mushrooms and onions). Don hates pepperoni. Sara loves mushrooms but will eat any of the toppings.

The slices are numbered from 1 to 12. Which slices should they take?
(*Note:* There is more than one possible solution.)

Bob: ______________ Don: ______________

Sara: ______________ Alice: ______________

Date Time

Math Boxes 6.2

1. Solve.

a. $1{,}000 * 204 =$ ______________

b. $10{,}000 * 6 =$ ______________

c. ______________ $= 940 * 1{,}000{,}000$

d. ______________ $= 320 * 100$

e. ______________ $= 76 * 100{,}000$

2. Estimate an answer for each problem.

a. $20.6 \div 4$ Estimate ______________

b. $184.38 \div 9$ Estimate ______________

c. $15.503 \div 7$ Estimate ______________

3. Mr. Hernandez's class took a survey to find out when students prefer to do their homework. They got responses from 125 fifth grade students. The results are shown in the table below.

As soon as I get home	17%
After having an after-school snack	30%
Right after dinner	39%
Just before going to bed	14%

Make a circle of the results, using your Geometry Template. Give the graph a title. Label the sections of the graph.

____

(title)

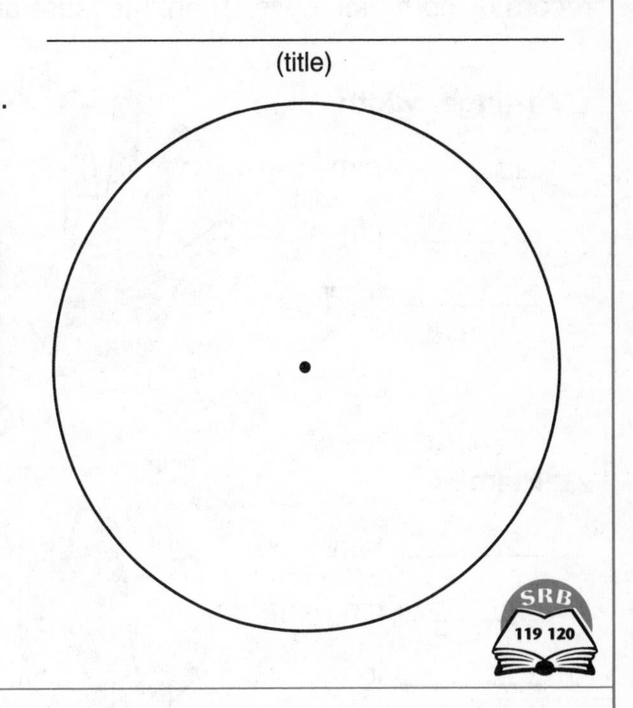

SRB 119 120

4. Write the following numbers in order from least to greatest.

$\frac{9}{2}$ 4.75 $\frac{13}{4}$ 4.8 $4\frac{7}{8}$

______, ______, ______, ______, ______

SRB 32 66 89

5. Rename each fraction as a decimal.

a. $\frac{15}{40} =$ ______________

b. $\frac{9}{12} =$ ______________

c. $\frac{2}{100} =$ ______________

d. $\frac{33}{99} =$ ______________

e. $\frac{40}{50} =$ ______________

Date Time

Personal Measures

Reference
10 millimeters (mm) = 1 centimeter (cm)
100 centimeters = 1 meter (m)
1,000 millimeters = 1 meter
1 inch (in.) is equal to about $2\frac{1}{2}$ (2.5) centimeters.

Work with a partner. You will need a ruler and a tape measure. Both tools should have both metric units (millimeters and centimeters) and U.S. customary units (inches).

Find your own personal measures for each body unit shown. First, measure and record using metric units. Then, measure and record using U.S. customary units.

1. **1-finger width**

   ________ mm

   ________ cm

   ________ in.

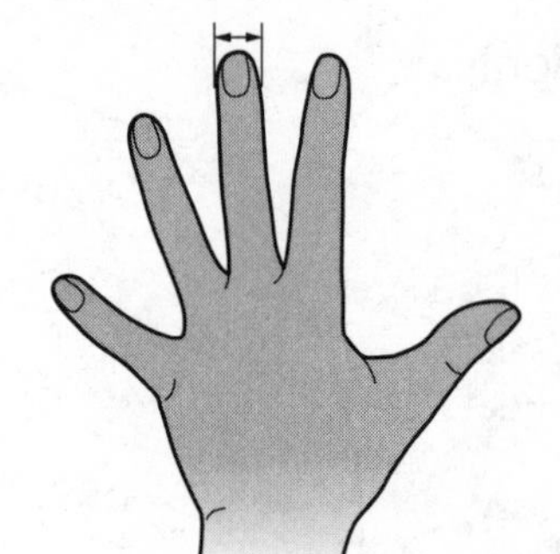

2. **Palm**

   ________ mm

   ________ cm

   ________ in.

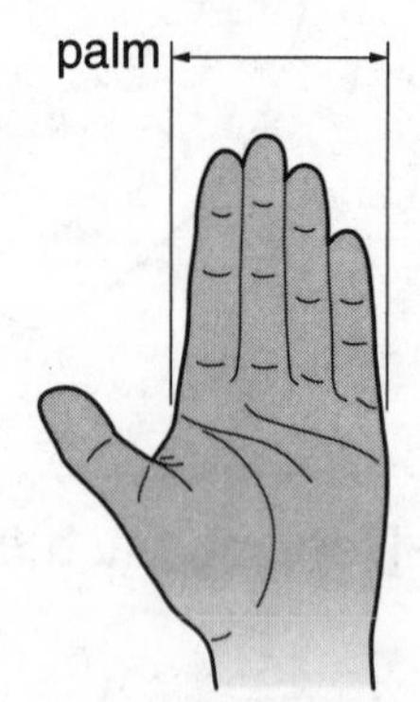

3. **Joint**

   ________ mm

   ________ cm

   ________ in.

joint

Date Time

Personal Measures (cont.)

4. Finger stretch

________ mm

________ cm

________ in.

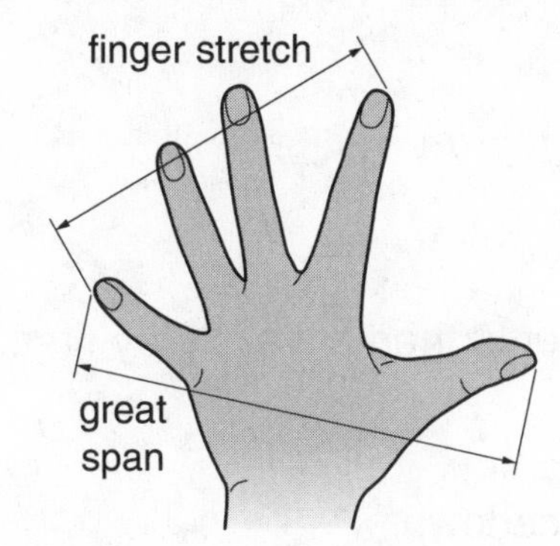

5. Great span

________ mm

________ cm

________ in.

6. Cubit

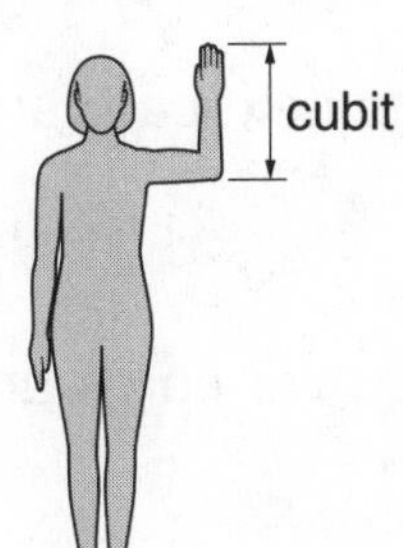
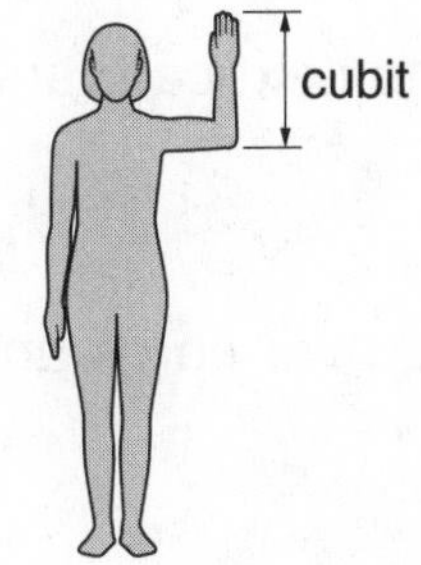

________ mm

________ cm

________ in.

7. Fathom

________ mm

________ cm

________ in.

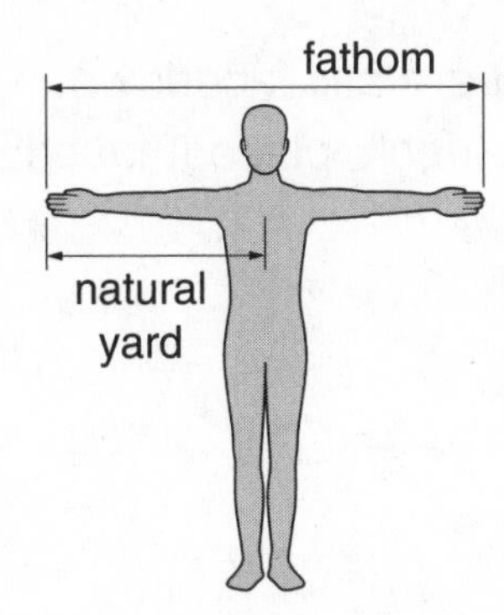

8. Natural yard

________ mm

________ cm

________ in.

Date Time

First to 21

Materials A deck of cards, consisting of four of each of the numbers 4, 5, 6, 7, and 8 (Do not use any other cards.)

Number of Players 2

Directions

Decide who will go first. That person should then always play first, whenever you start a new game.

1. Shuffle the cards. Place the deck facedown.

2. The player going first turns over the top card and announces its value.

3. The player going second turns over the next card and announces the total value of the two cards turned over.

4. Partners continue to take turns turning over cards and announcing the total value of all the cards turned over so far.

5. The winner is the first player to correctly announce "21" or any number greater than 21.

6. Start a new game using the cards that are still facedown. If all of the cards are turned over during a game, shuffle the deck, place it facedown, and continue.

Estimation Challenge

A **fair game** is one that each player has the same chance of winning. If there is an advantage or disadvantage in playing first, then the game is not fair.

With your partner, investigate whether *First to 21* is a fair game.

Collect data by playing the game.
Over the next week, play *First to 21* at least 50 times. Keep a tally each day. Show how many times the player going first wins, and how many times the player going second wins.

Date	Player Going First Wins	Player Going Second Wins	Total Games to Date

Enter your results on the classroom tally sheet.
Each day you play the game, record the results on the tally sheet for the whole class that your teacher has prepared.

Each day you play, ask yourself:

- What is my estimate for the chance that the player going first will win?
- What is my estimate for the chance that the player going second will win?
- Do my estimates change as more and more games are played?
- Does *First to 21* seem to be a fair game?

Date Time

Hand and Finger Measures: The Great Span

For measurements on this page and the next page:

If you are right-handed, measure your left hand.
If you are left-handed, measure your right hand.

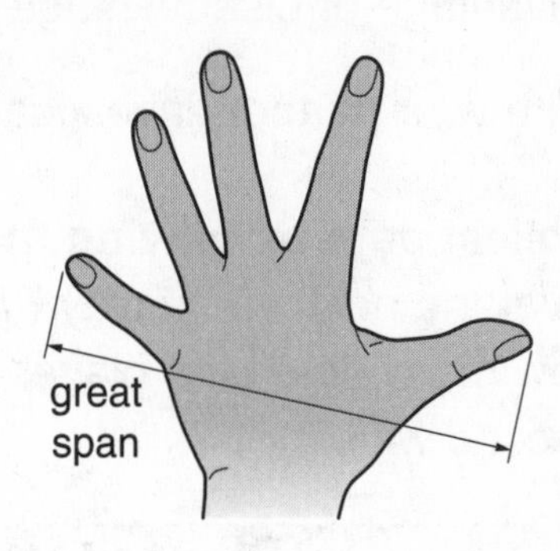

Your **great span** is the distance from the tip of your thumb to the tip of your little finger. Place the tip of your thumb at the bottom of the ruler in the margin (at 0). Extend your fingers. Stretch your little finger as far along the ruler as you can. Read your great span to the nearest millimeter and record it below.

My great span is about ________ millimeters.

Your teacher will show you how to use the table below. Use it to record the great-span data for your class. The result is called a **stem-and-leaf plot.**

Great-Span Measurements for the Class (millimeters)

Stems (100s and 10s)	**Leaves** (1s)
13	________
14	________
15	________
16	________
17	________
18	________
19	________
20	________
21	________
22	________
23	________
24	________

Landmarks for the class great-span data:

Minimum: ________ mm

Maximum: ________ mm

Mode(s): ________ mm

Median: ________ mm

cm 1 2 3 4 5 6 7 8 9 10 11 12 13 14 15 16 17 18 19 20 21 22 23 24

Hand and Finger Measures: Finger Flexibility

A measure of finger flexibility is how far apart you can spread your fingers. The picture shows how to measure the **angle of separation** between your thumb and first (index) finger.

1. Spread your thumb and first finger as far apart as you can. Do this in the air. Don't use your other hand to help. Lower your hand onto a sheet of paper. Trace around your thumb and first finger. With a straightedge, draw two line segments to make a V shape, or angle, that fits the finger opening. Use a protractor to measure the angle between your thumb and first finger. Record the measure of the angle.

Measure this angle.

Angle formed by **thumb** and **first** finger:

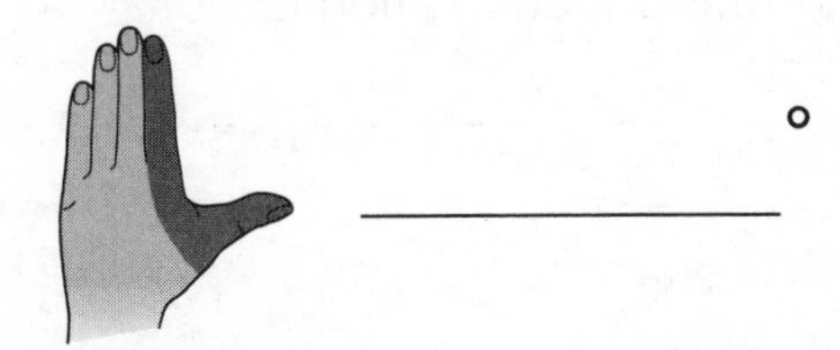

______________ °

2. In the air, spread your first and second fingers as far apart as possible. On a sheet of paper, trace these fingers and draw the angle of separation between them. Measure the angle and record its measure.

Angle formed by **first** and **second** fingers:

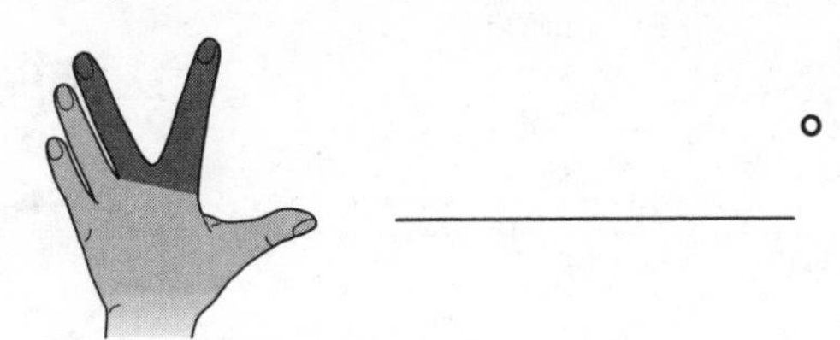

______________ °

3. Record the class landmarks for both finger-separation angles in the table at the right.

Landmark	Thumb and First	First and Second
Minimum		
Maximum		
Mode(s)		
Median		

Date Time

Math Boxes 6.3

1. When Antoinette woke up on New Year's Day, it was $-4°F$ outside. By the time the parade started, it was a cozy 18°F. How many degrees had the temperature risen by the time the parade began?

____

2. Write each numeral in number-and-word notation.

a. 43,000,000 _______________

b. 607,000 _______________

c. 3,000,000,000 _______________

d. 72,000 _______________

3. Circle the name(s) of the shape(s) that could be partially hidden behind the wall.

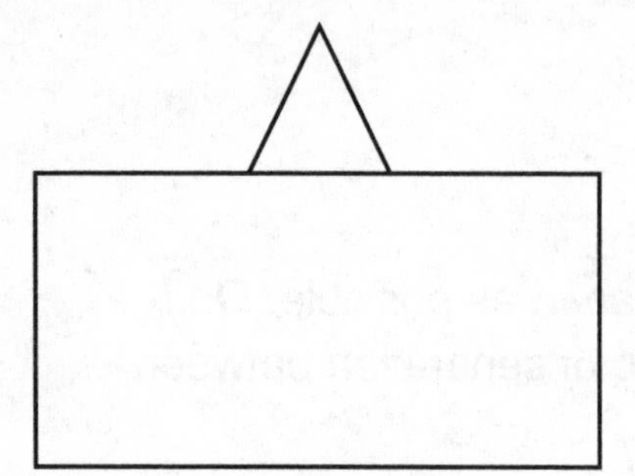

rectangle pentagon rhombus

4. Write the prime factorization of 80.

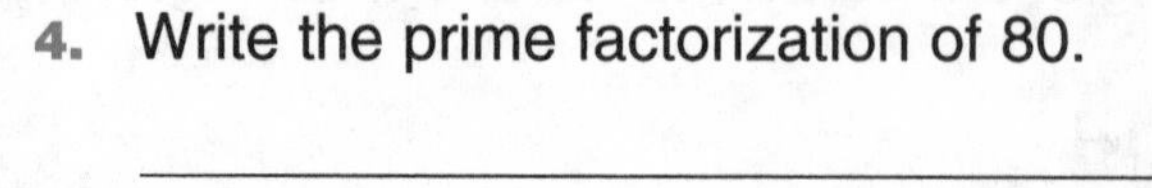

5. Write a number story for $37 * 68$. Then solve it.

____

____

____

____

____

____

Answer: _______________

Date Time

Math Boxes 6.4

1. Write a 10-digit numeral that has
   7 in the billions place,
   5 in the hundred-thousands place,
   3 in the ten-millions place,
   4 in the tens place,
   8 in the hundreds place, and
   2 in all other places.

   ____ , ____ ____ ____ , ____ ____ ____ , ____ ____ ____

   Write the numeral in words.

   ____

   ____

2. Round each number to the nearest whole number.

   a. 36.084 ________
   b. 25.9 ________
   c. 63.52 ________
   d. 70.364 ________
   e. 89.7 ________

3. Write a fraction or a mixed number for each of the following.

   a. 5 minutes = ________ hour
   b. 20 minutes = ________ hour
   c. 35 minutes = ________ hour
   d. 55 minutes = ________ hour
   e. 10 minutes = ________ hour

4. Rename each mixed number as a fraction.

   a. $3\frac{7}{8}$ = ________
   b. $4\frac{6}{9}$ = ________
   c. $10\frac{7}{12}$ = ________
   d. $8\frac{2}{3}$ = ________
   e. $6\frac{5}{14}$ = ________

5. Complete.

   a. ____________ * 20 = 6,000
   b. ____________ = 800 * 40
   c. ____________ * 600 = 30,000
   d. ____________ * 50 = 25,000
   e. 54,000 = 60 * ____________

Date Time

Mystery Plots

There are five line plots on page 183. Each plot shows a different set of data about a fifth grade class.

Match each of the following four sets of data with one of the five plots. Then fill in the "Unit" for each matched graph on page 183.

1. The number of hours of TV each fifth grader watched last night Plot ________
2. The ages of the younger brothers and sisters of the fifth graders Plot ________
3. The heights, in inches, of some fifth graders Plot ________
4. The ages of some fifth graders' grandmothers Plot ________
5. Explain how you selected the line plot for Data Set 4.

____

____

____

____

____

6. Tell why you think the other line plots are not correct for Data Set 4.

____

____

____

____

____

____

Date Time

Mystery Plots (cont.)

Plot #1 Unit: ____________________

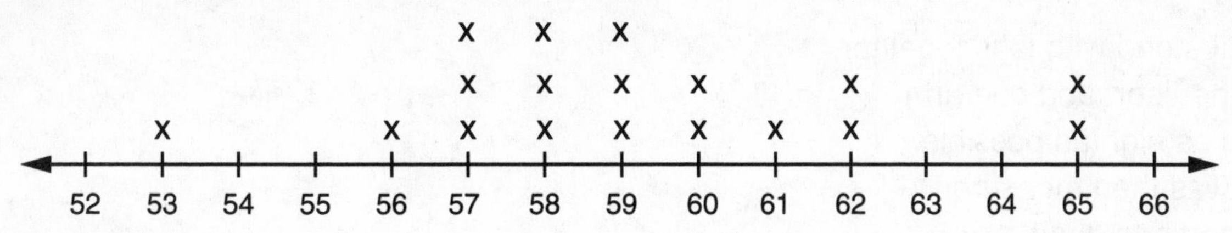

Plot #2 Unit: ____________________

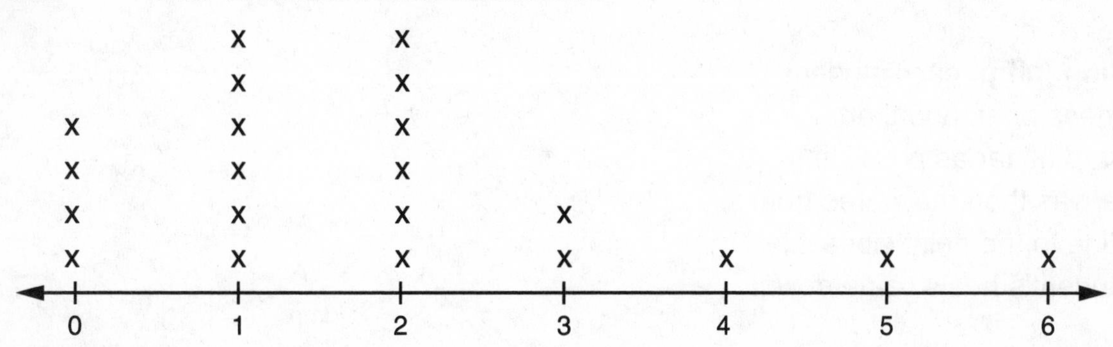

Plot #3 Unit: ____________________

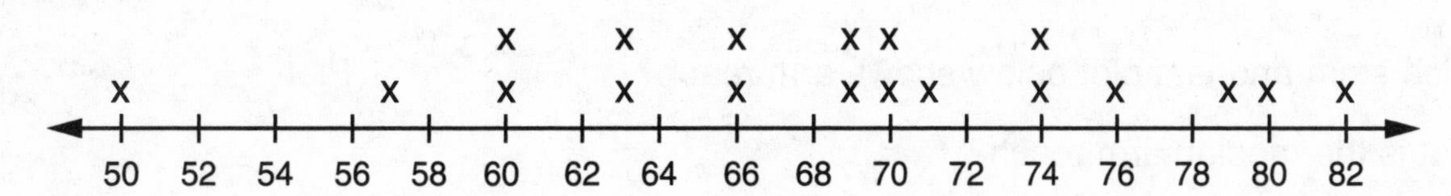

Plot #4 Unit: ____________________

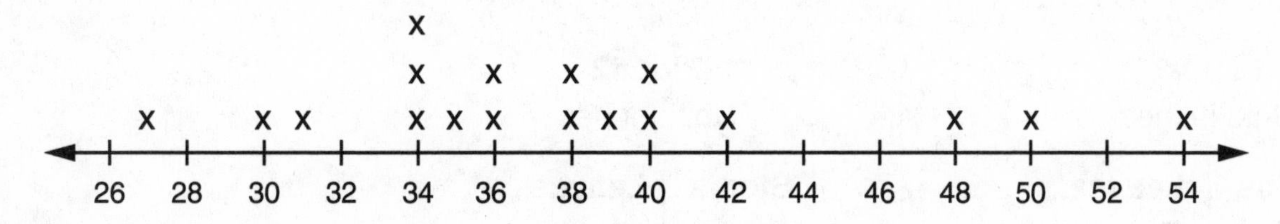

Plot #5 Unit: ____________________

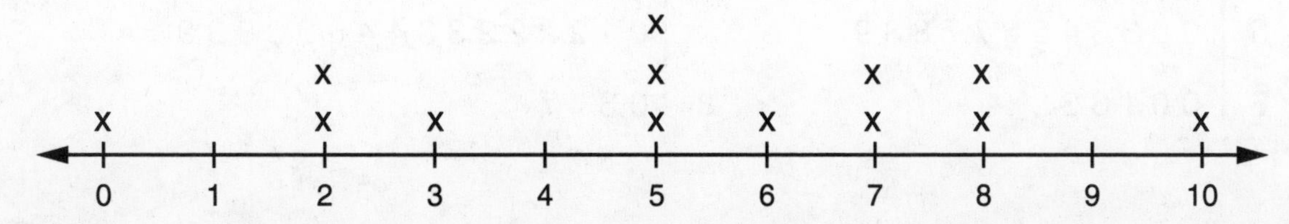

Date Time

Reaching and Jumping

Students in a fifth grade class measured how far they each could reach and jump.

Each student stood with legs together, feet flat on the floor, and one arm stretched up as high as possible. **Arm reach** was then measured from top fingertip to floor.

arm reach

In the **standing jump,** each student stood with knees bent, and then jumped forward as far as possible. The distance was then measured from the starting line to the point closest to where the student's heels came down.

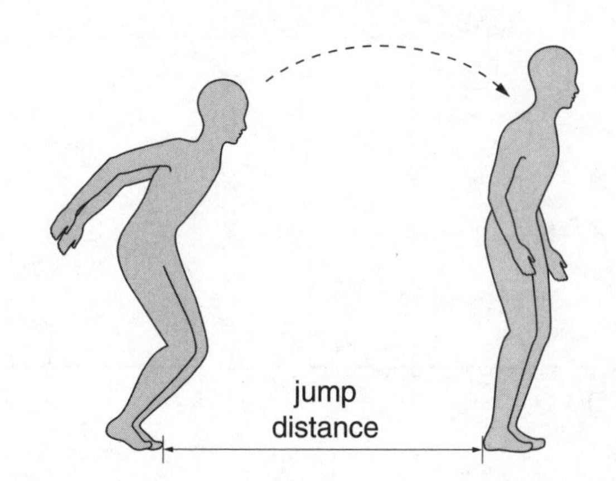

The students made stem-and-leaf plots of the results.

1. a. Which stem-and-leaf plot below shows arm reach? Plot ________

   b. What is the median arm reach? ________ in.

2. a. Which stem-and-leaf plot below shows standing-jump distances? Plot ________

   b. What is the median standing-jump distance? ________ in.

Plot #1
Unit: inches

Stems (10s)	**Leaves** (1s)
4	4 6 8
5	0 0 3 3 4 5 6 7 7 8 8 9
6	0 0 1 3 3 8

Plot #2
Unit: inches

Stems (10s)	**Leaves** (1s)
6	7
7	0 1 2 2 2 2 3 3 4 4 6 6 6 8 9 9
8	0 3 4 7

Sampling Candy Colors

1. You and your partner each take 5 pieces of candy from the bowl. Combine your candies and record your results in the table under Our Sample of 10 Candies.

Candy Color	Our Sample of 10 Candies		Combined Class Sample	
	Count	Percent	Count	Percent

2. Your class will work together to make a sample of 100 candies. Record the counts and percents of the class sample under Combined Class Sample in the table.

3. Finally, your class will count the total number of candies in the bowl and the number of each color.

   a. How well did your sample of 10 candies predict the number of each color in the bowl? ______

   b. How well did the combined class sample predict the number of each color in the bowl? ______

   c. Do you think that a larger sample is more trustworthy than a smaller sample? ______

   Explain your answer. ______

Date Time

Solving Part-Whole Fraction Problems

1. How much is $\frac{3}{5}$ of \$1? ________

2. How much is $\frac{3}{5}$ of \$10? ________

3. How much is $\frac{3}{5}$ of \$1,000? ________

4. Eight counters is $\frac{1}{2}$ of the set. How many counters are in the set? ________ counters

5. Twenty counters is $\frac{2}{10}$ of the set. How many counters are in the set? ________ counters

6. A set has 40 counters. How many counters are in $\frac{3}{8}$ of the set? ________ counters

7. A set has 36 counters. How many counters are in $\frac{5}{6}$ of the set? ________ counters

8. Mariah shared her sandwich equally with her 3 friends. What fraction of a sandwich did Mariah get? ________ of a sandwich

9. Bernice gave $\frac{2}{3}$ of her 18 fancy pencils to her best friend. How many pencils did Bernice have left? ________ pencils

Challenge

10. Jamie and his two friends shared $\frac{1}{2}$ of his 12 candies. How many candies did each friend get? ________ candies

11. Explain how you solved Problem 10. ____

____

____

____

Date Time

Math Boxes 6.5

1. Solve.

a. 100,000 ∗ 300 = ____________

b. 100 ∗ 5,060 = ____________

c. ____________ = 728 ∗ 10,000

d. ____________ = 6,434 ∗ 1,000

e. ____________ = 120 ∗ 10,000

2. Estimate an answer for each problem.

a. $4\overline{)39.04}$ Estimate ________

b. $8\overline{)17.6}$ Estimate ________

c. $5\overline{)300.007}$ Estimate ________

3. Draw a circle graph that is divided into the following sectors: 32%, 4%, 22%, 18%, and 24%. Make up a situation for the graph. Give the graph a title. Label each section.

Description:

____

____

____

____
(title)

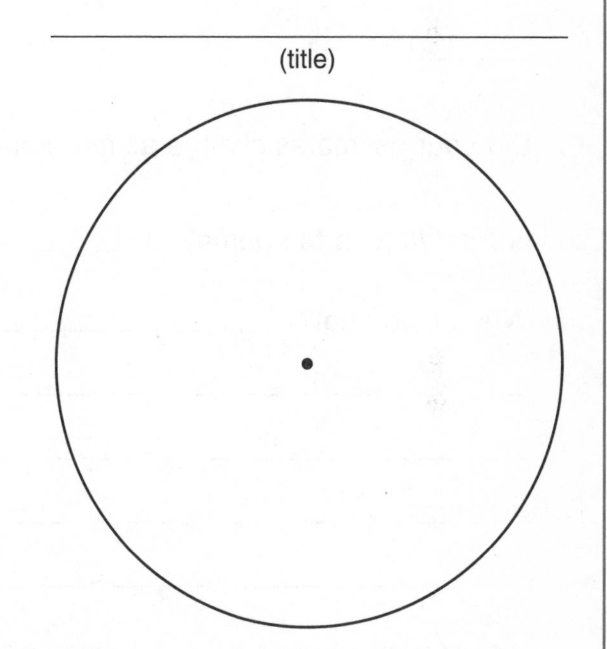

4. Write the following numbers in order from least to greatest.

5.03 $4\frac{7}{4}$ 5.3 $\frac{3}{15}$ $5\frac{2}{5}$

______, ______, ______, ______, ______

5. Rename each fraction as a decimal.

a. $\frac{24}{36}$ = ________

b. $\frac{78}{100}$ = ________

c. $\frac{25}{40}$ = ________

d. $\frac{10}{15}$ = ________

e. $\frac{21}{28}$ = ________

Date Time

Is *First to 21* a Fair Game?

1. What is the total number of *First to 21* games your class has played?

   _________ games

2. How many games did the player going first win? _______ games

3. How many games did the player going second win? _______ games

4. What is your best estimate for the chance that the player going first will win?

   ____

5. What is your best estimate for the chance that the player going second will win?

   ____

6. Did your estimates change as more and more games were played? _______

7. Is *First to 21* a fair game? _______

   Why or why not? ____

   ____

   ____

   ____

   ____

   If *First to 21* isn't a fair game, how could you make it more fair?

   ____

   ____

   ____

   ____

   ____

Date Time

Math Boxes 6.6

1. The temperature in Chicago at 6 P.M. was 35°F. By midnight, the temperature had dropped 48 degrees. What was the temperature at midnight?

   ____

2. Write each numeral in number-and-word notation.

   a. 56,000,000 ______________

   b. 423,000 ______________

   c. 18,000,000,000 ______________

   d. 9,500,000 ______________

3. What kind of regular polygon could be partially hidden behind the wall?

   ____

   Complete the shape.

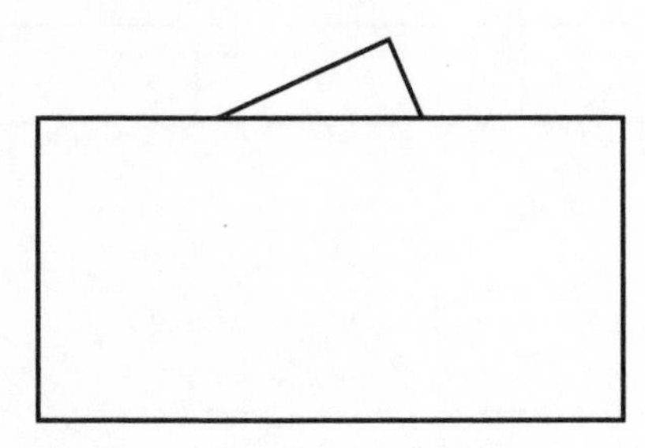

4. Write the prime factorization of 132.

   ____

5. Write a number story for 81 * 17. Then solve it.

   ____

   ____

   ____

   ____

   ____

   ____

   Answer: ______________

Date Time

Frequency Tables

A **frequency table** is a chart on which data is tallied to find the frequency of given events or values.

Use the frequency tables below to tally the Entertainment data and Favorite-Sports data on page 110 in your *Student Reference Book.* Then complete the tables. If you conducted your own survey, use the frequency tables to tally the data you collected. Then complete the tables.

1. What is the survey question? ______________________________

____

Category	Tallies	Number	Fraction	Percent

Total number of tallies ________

2. What is the survey question? ______________________________

____

Category	Tallies	Number	Fraction	Percent

Total number of tallies ________

Use with Lesson 6.6.

Date Time

Data Graphs and Plots

1. Draw a bar graph for one of the survey questions on journal page 190. Label the parts of the graph. Give the graph a title.

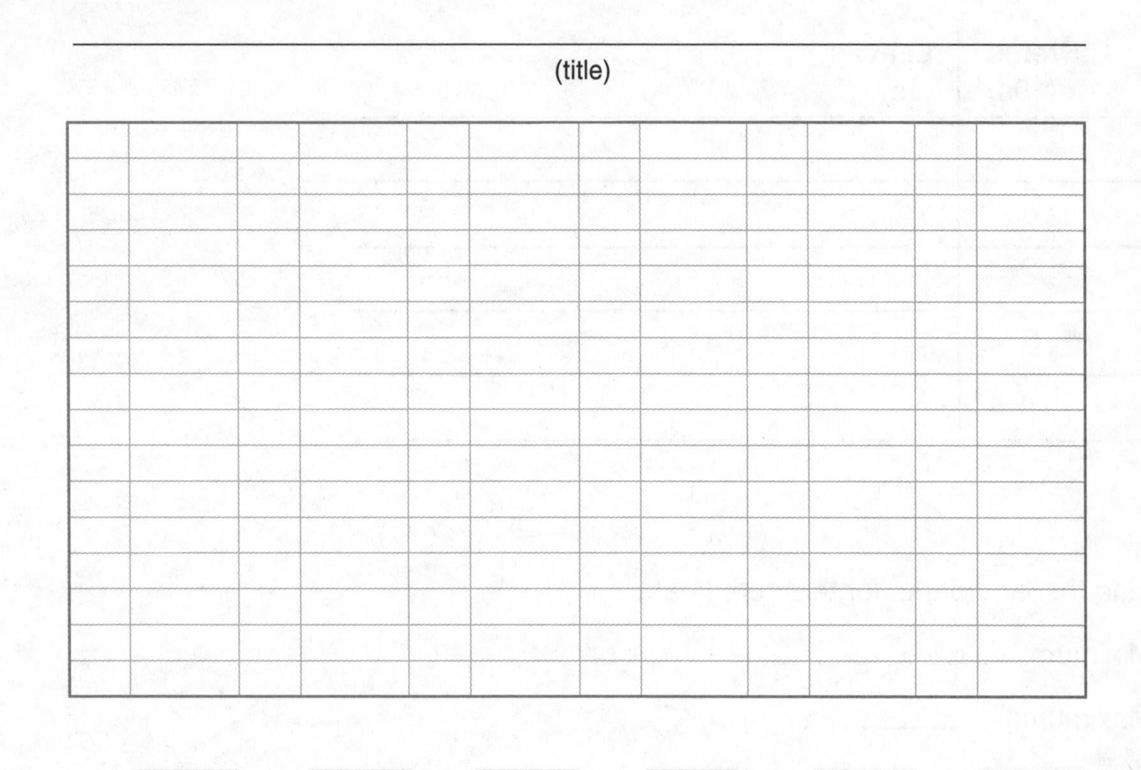

2. Draw a circle graph for the other survey question on journal page 190. Label the sections of the graph. Give the graph a title.

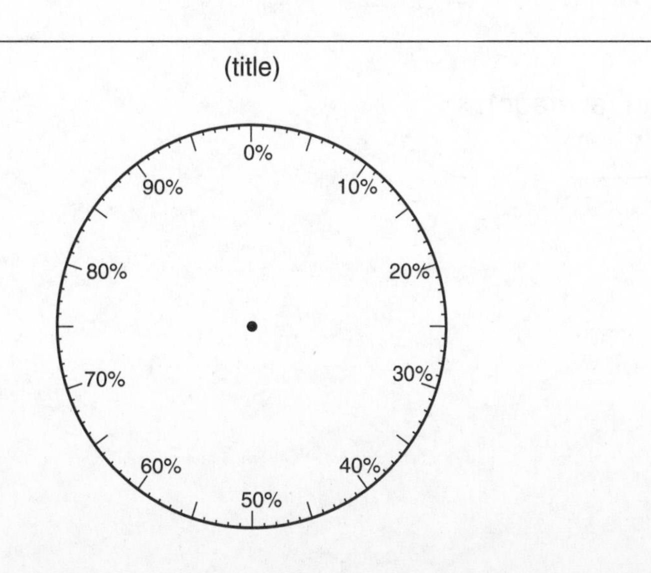

Data Graphs and Plots (cont.)

3. Make a stem-and-leaf plot for the Shower/Bath Time data on page 110 in your *Student Reference Book*. If you conducted your own survey, make a stem-and-leaf plot for the data you collected.

Stems (10s)	**Leaves** (1s)

Find the landmarks for this set of data.

Minimum: ____________

Maximum: ____________

Range: ____________

Median: ____________

Mode: ____________

Challenge

4. Calculate the mean (average).

Mean: ____________

Date Time

Place-Value Puzzles

1. For each problem, tell what you would enter in your calculator to change the numbers.

Starting Number	Ending Number	Calculator Key Strokes
34,728	34,758	
1,176	276	
62,885	71,885	
109,784	110,084	
9,002	8,996	

2. Use the clues to write the 7-digit number.
The digit in the tens place is 7.
The digit in the hundred-thousands place is 6 less than double 7.
The digit in the ones place is $\frac{1}{4}$ of three times the digit in the hundred-thousands place.
The digit in the other places is the smallest even digit.

_____ , _____ _____ _____ , _____ _____ _____

Challenge

3. Use the clues to write the 4-digit number.
The digit in the tens place is double the digit in the tenths place.
The digit in the hundredths place is $\frac{1}{2}$ the digit in the tenths place.
The digit in the ones place is the only 8 in the number.

_____ _____ . _____ _____

4. I am a two-digit composite number. One of my digits is worth three times as much as the other digit. Double me is less than 100. I am not divisible by 3. What am I?

_____

Date Time

Climate Maps

To answer the questions below, use the "Average Yearly Precipitation in the U.S." and "Growing Seasons in the U.S." maps on page 338 of your American Tour.

The precipitation map shows the average amount of moisture that falls as rain and snow in one year. Snow is translated into an equivalent amount of rain.

The growing seasons map shows the average number of months between the last frost in spring and the first frost in fall. During this time, the temperature remains above freezing (32°F or 0°C), and crops may be grown.

1. Denver, Colorado, receives about _____________ inches of precipitation as rain and snow per year.

   Denver's growing season is about _____________ months long.

2. Los Angeles, California, receives about _____________ inches of precipitation per year.

   The growing season in Los Angeles is _____________ months long.

3. a. According to these maps, how are Los Angeles and New Orleans similar?

   ____

   ____

   ____

   b. Who is more likely to be worried about a lack of rain: a farmer near Los Angeles or a farmer near New Orleans? Why?

   ____

   ____

   ____

   ____

Climate Maps (cont.)

4. In general, does it rain more in the eastern states or in the western states?

____

5. In general, is the growing season longer in the northern states or in the southern states? _______________

6. Cotton needs a growing season of at least 6 months. In the list below, circle the states most likely to grow cotton.

Texas Nebraska Mississippi Ohio

7. North Dakota and Kansas are the largest wheat-producing states.

What is the length of the growing season in North Dakota? _______________

What is the length of the growing season in Kansas? _______________

About how much precipitation does North Dakota receive per year? _______________

About how much precipitation does eastern Kansas receive per year? _______________

8. a. Locate the Rocky Mountains on your landform map (American Tour, page 339). What is the growing season for this mountain area?

____

____

b. What is the growing season for the Appalachian Mountains area?

____

____

Number Stories

1. Brenda bought 4 cheeseburgers for her family for lunch. The total cost was $5.56.

   How much did 2 cheeseburgers cost? ____________

2. Thomas's family went on a long trip over summer vacation. They drove for 5 days. The distances for the 5 days were as follows: 347 miles, 504 miles, 393 miles, 422 miles, and 418 miles.

   a. To the nearest mile, what was the average distance traveled per day? ____________

   b. Tell what you did with the remainder. Explain why. ____________

   ____

   ____

3. Justin's school has 15 classrooms. On an average, there are 28 students per room. One fifth of the classrooms are for fifth graders. About how many students are in the school? ____________

4. Carolyn reads 45 pages of a book every night. How many pages did she read in the month of March (31 days)? ____________

5. Lucienne and her class made 684 notecards for a school benefit.

   a. How many boxes of eight can they fill? ____________

   b. Explain what the remainder represents and what you did with it. ____________

   ____

   ____

   ____

   ____

Date Time

Math Boxes 6.7

1. Subtract. (*Hint:* Use a number line to help you.)

   a. 8 − 15 = ________

   b. 16 − 18 = ________

   c. ________ = 47 − 51

   d. ________ = 30 − 24

   e. ________ = 32 − 29

2. Rewrite each number in expanded notation.

   a. 3^4 = 3 * 3 * 3 * 3

   b. 5^3 = ________________

   c. 7^4 = ________________

   d. 2^5 = ________________

   e. 10^3 = ________________

3. Below are the distances (in feet) a baseball must travel to right field in order to be a home run in various major-league baseball parks. Circle the stem-and-leaf plot below that represents this data.

   330, 353, 330, 345, 325, 330, 325, 338, 318,

   302, 333, 347, 325, 315, 330, 327, 314, 348

Stems (100s and 10s)	Leaves (1s)
30	0 2 5
31	0 0 8
32	5 5 5 5 5
33	0 0 8 8 8
34	5 7
35	3
36	1

Stems (100s and 10s)	Leaves (1s)
30	2
31	4 5 8
32	5 7
33	0 3 8
34	5 7 8
35	3
36	

Stems (100s and 10s)	Leaves (1s)
30	2
31	4 5 8
32	5 5 5 7
33	0 0 0 0 3 8
34	5 7 8
35	3
36	

Date Time

Adding and Subtracting Fractions on a Slide Rule

Use your slide rule, or any other method, to add or subtract.

1. $\frac{1}{2} + \frac{1}{4} =$ ______________

2. $\frac{5}{8} + \frac{2}{8} =$ ______________

3. $2\frac{1}{2} + 3 =$ ______________

4. $3\frac{5}{8} + 3\frac{3}{4} =$ ______________

5. $1\frac{9}{16} + 1\frac{5}{16} =$ ______________

6. $\frac{7}{8} - \frac{3}{8} =$ ______________

7. $5\frac{3}{4} - 2\frac{1}{4} =$ ______________

8. $7\frac{1}{2} - 4\frac{5}{8} =$ ______________

9. $\frac{19}{16} - \frac{1}{2} =$ ______________

10. $5\frac{1}{2} - 6 =$ ______________

11. Put a star next to the problems above that you thought were the easiest.

12. Complete the following:

It is easy to add or subtract fractions with the same denominator (for example, $\frac{4}{8} - \frac{3}{8}$)

because ____

____

____

____

Prime Time

When this book went to the printer, the largest known prime number was equal to $2^{6,972,593} - 1$, a number with 2,098,960 digits. If these digits were printed on one line, 6 digits to a centimeter, they would stretch almost 3.5 kilometers. Checking that this number is prime took 111 days of part-time work by a desktop computer. The person who found it qualifies for a prize of $50,000, offered by the Electronic Frontier Foundation. A prize of $100,000 is being offered to the first person who finds a prime number with at least 10 million digits.

Large prime numbers are used in writing codes and testing computer hardware. More about the search for prime numbers can be found on the Internet at http://www.mersenne.org/ and http://ontko.com/~rayo/primes.index.html.

Date Time

Adding and Subtracting Fractions with Fraction Sticks

Write the missing fraction for each pair of fraction sticks. Then write the sum or difference of the fractions.

1. $\frac{5}{12}$ + ________ = ____________

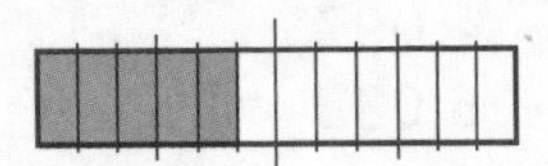

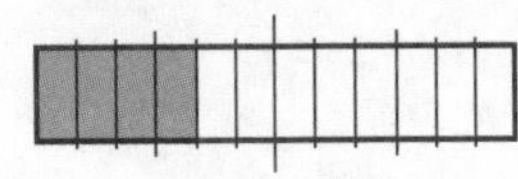

2. $\frac{5}{6}$ − ________ = ____________

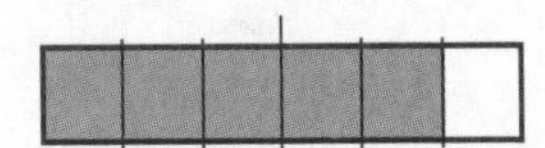

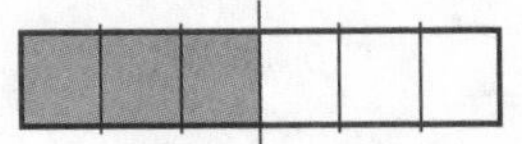

3. ________ − $\frac{1}{4}$ = ____________

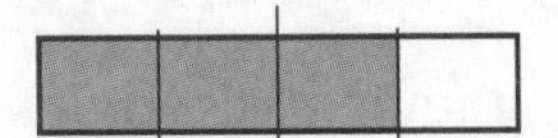

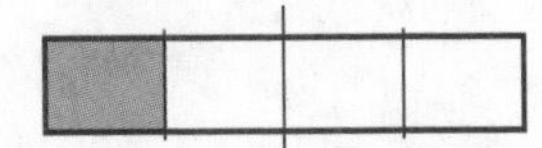

4. Andy jogs on a track where each lap is $\frac{1}{4}$ mile. Find the number of miles he jogged each day and then the total number of laps and miles for the three days.

Day	Laps	Distance
Monday	5	
Wednesday	10	
Thursday	8	
Total		

Date Time

Math Boxes 6.8

1. Solve.

a. $43 + 82$

b. $4{,}097 + 6{,}035$

c. $47 - 18$

d. $624 - 575$

e. $503 - 426$

2. The bar graph shows the favorite flavors of ice cream of Mr. Lenard's fifth grade students.

a. How many students prefer the class's favorite flavor? ________

b. How many more students prefer chocolate than vanilla? ________

c. Mark your favorite flavor with an X.

Fifth Grade Favorite Ice Cream Flavors

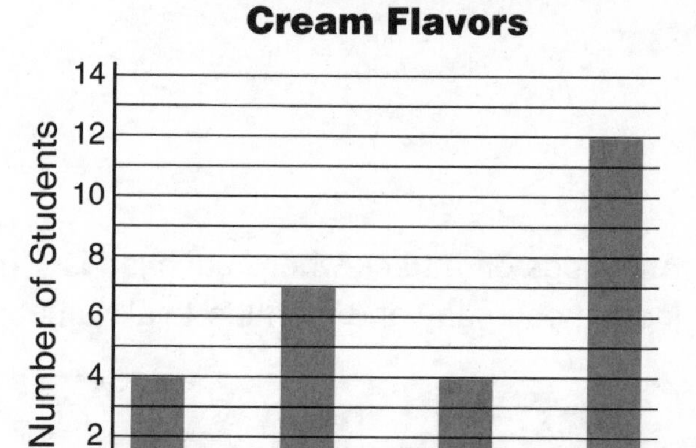

Vanilla Chocolate Strawberry Chocolate Chip

SRB 116

3. Use your compass and the map scale to estimate the distance from the hotel to the museum shown on the map.

The distance is about ______________.

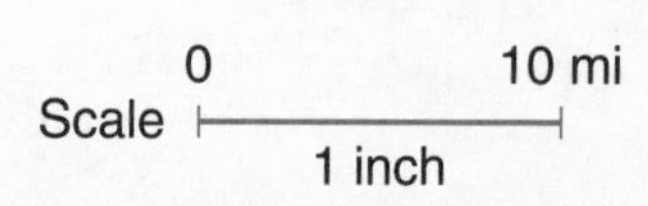

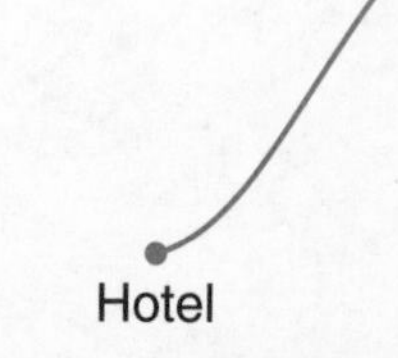

Hotel

Museum

SRB 195 196

Date Time

Clock Fractions

Part 1: Math Message

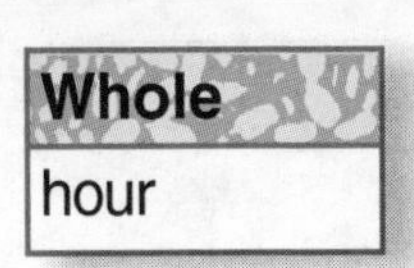

The numbers on a clock face divide one hour into twelfths. Each $\frac{1}{12}$ of an hour is 5 minutes.

How many minutes does each of the following fractions and mixed numbers represent? The first one has been done for you.

1. $\frac{1}{12}$ hr = 5 min **2.** $\frac{5}{12}$ hr = ________ min **3.** $\frac{1}{2}$ hr = ________ min

4. $\frac{1}{3}$ hr = ________ min **5.** $\frac{1}{4}$ hr = ________ min **6.** $\frac{1}{6}$ hr = ________ min

Part 2

Using the clock face, fill in the missing numbers. The first one has been done for you.

7. $\frac{1}{4}$ hr = $\frac{3}{12}$ hr **8.** $\frac{8}{12}$ hr = $\frac{2}{\square}$ hr **9.** $\frac{1}{3}$ hr = $\frac{2}{\square}$ hr

10. $\frac{\square}{12}$ hr = $\frac{5}{6}$ hr **11.** $\frac{3}{\square}$ hr = $\frac{9}{12}$ hr **12.** $\frac{2}{12}$ hr = $\frac{\square}{6}$ hr

13. $1\frac{1}{2}$ hr = $\frac{\square}{4}$ hr **14.** $\frac{5}{3}$ hr = $\frac{\square}{12}$ hr **15.** $\frac{4}{12}$ hr = $\frac{1}{\square}$ hr

Part 3

Use clock fractions, if helpful, to solve these problems. Write each answer as a fraction.

Example $\frac{3}{4} - \frac{1}{3} = ?$

Think: 45 minutes − 20 minutes = 25 minutes

So $\frac{3}{4} - \frac{1}{3} = \frac{5}{12}$

16. $\frac{5}{12} + \frac{3}{12} =$ ________ **17.** $\frac{3}{4} + \frac{2}{4} =$ ________ **18.** $\frac{11}{12} - \frac{3}{12} =$ ________

19. $1 - \frac{2}{3} =$ ________ **20.** $\frac{5}{4} - \frac{2}{4} =$ ________ **21.** $\frac{2}{3} + \frac{1}{6} =$ ________

22. $\frac{1}{4} + \frac{1}{3} =$ ________ **23.** $\frac{1}{3} - \frac{1}{4} =$ ________ **24.** $\frac{5}{6} - \frac{3}{4} =$ ________

Date Time

Using a Common Denominator

Study the examples. Then work the problems below in the same way.

Example 1 $\frac{2}{3} + \frac{1}{6} = ?$			***Example 2*** $\frac{5}{6} - \frac{3}{4} = ?$		
Unlike Denominators		Common Denominators	Unlike Denominators		Common Denominators
$\frac{2}{3}$	$\frac{2}{3} = \frac{4}{6}$	$\frac{4}{6}$	$\frac{5}{6}$	$\frac{5}{6} = \frac{10}{12}$	$\frac{10}{12}$
$+\ \frac{1}{6}$		$+\ \frac{1}{6}$	$-\ \frac{3}{4}$	$\frac{3}{4} = \frac{9}{12}$	$-\ \frac{9}{12}$
		$\frac{5}{6}$			$\frac{1}{12}$

1. $\frac{2}{3} + \frac{2}{9} = ?$

Unlike Denominators

$$\frac{2}{3} + \frac{2}{9}$$

Common Denominators

2. $\frac{13}{16} - \frac{3}{4} = ?$

Unlike Denominators

$$\frac{13}{16} - \frac{3}{4}$$

Common Denominators

3. $\frac{1}{3} + \frac{2}{5} = ?$

Unlike Denominators

$$\frac{1}{3} + \frac{2}{5}$$

Common Denominators

4. $\frac{5}{6} - \frac{4}{9} = ?$

Unlike Denominators

$$\frac{5}{6} - \frac{4}{9}$$

Common Denominators

Date Time

Using a Common Denominator (cont.)

5. $\frac{12}{4} + \frac{3}{2} = ?$

Unlike Denominators	Common Denominators
$\frac{12}{4}$ $+ \frac{3}{2}$ ______	

6. $1\frac{1}{16} - \frac{3}{8} = ?$

Unlike Denominators	Common Denominators
$1\frac{1}{16}$ $- \frac{3}{8}$ ______	

7. A piece of ribbon is $7\frac{1}{2}$ inches long. If a piece $2\frac{3}{16}$ inches long is cut off, how long is the remaining piece? ______________ in.

Write a number sentence to show how you solved the problem.

____

8. Three boards are glued together. The diagram below shows the thickness of each board. What is the total thickness of the three boards? ______________ in.

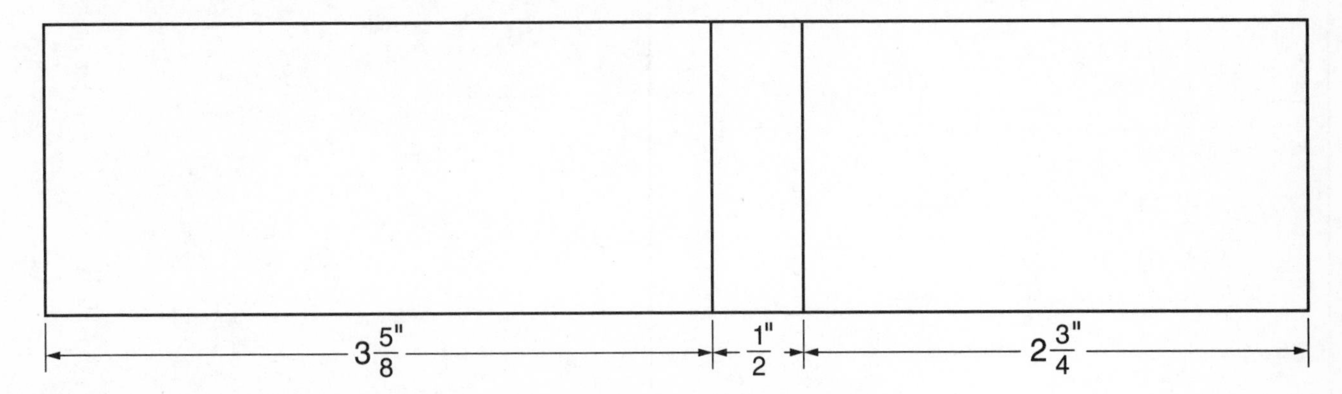

Write a number sentence to show how you solved the problem.

____

Date Time

Math Boxes 6.9

1. Subtract. (*Hint:* Use a number line to help you.)

   a. 50 − 56 = ________

   b. 48 − 68 = ________

   c. ________ = 23 − 29

   d. ________ = 99 − 105

   e. ________ = 75 − 73

2. Rewrite each number in exponential notation.

   a. 4 * 4 * 4 = ________

   b. 5 * 5 * 5 * 5 = ________

   c. 9 * 9 * 9 * 9 = ________

   d. 7 * 7 = ________

   e. 2 * 2 * 2 * 2 * 2 = ________

3. a. Make a stem-and-leaf plot for the bowling scores from the Pick's family reunion bowl.

   106, 135, 168, 162, 130, 116, 109, 139, 161,

   130, 118, 105, 150, 164, 130, 138, 112, 116

Stems (100s and 10s)	Leaves (1s)

   b. What is the maximum score? ________

   c. What is the mode for the scores? ________

   d. What is the median score? ________

Another Way to Find a Common Denominator

1. a. Draw a horizontal line to split each part of this thirds fraction stick into 2 equal parts. How many parts are there in all? ________

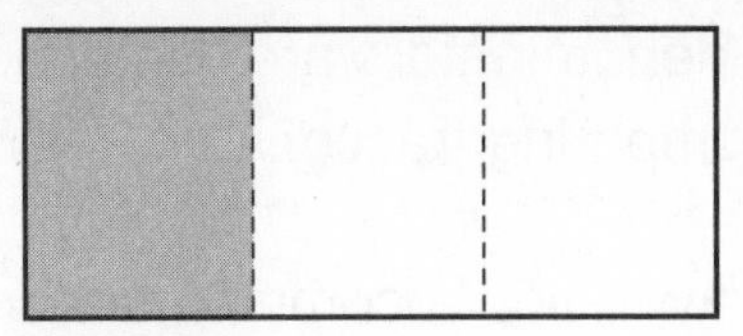

b. Draw horizontal lines to split each part of this halves fraction stick into 3 equal parts. How many parts are there in all? ________

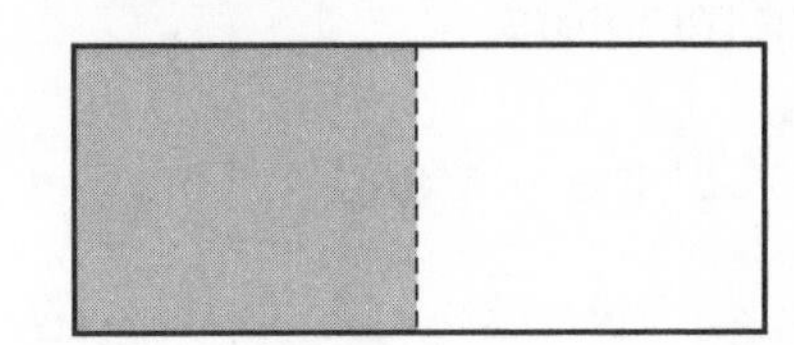

c. 

$\frac{\square * 1}{\square * 3} = \frac{2}{6}$ $\qquad$ $\frac{\square * 1}{\square * 2} = \frac{3}{6}$

2. a. If you drew lines to split each part of this fourths fraction stick into 6 equal parts, how many parts would there be in all? ________

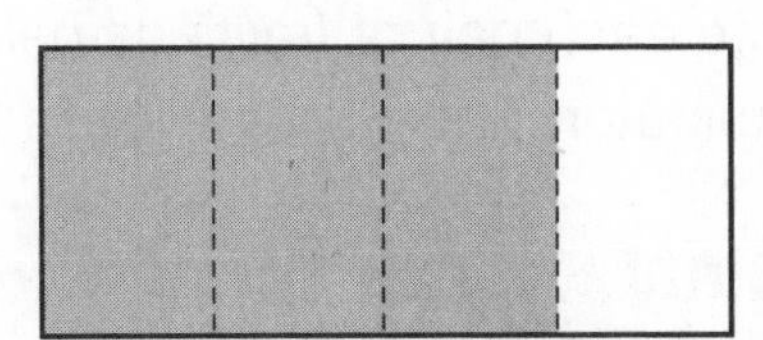

b. If you drew lines to split each part of this sixths fraction stick into 4 equal parts, how many parts would there be in all? ________

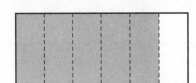

c. $\frac{\square * 3}{\square * 4} = \frac{18}{24}$ $\qquad$ $\frac{\square * 5}{\square * 6} = \frac{20}{24}$

3. One way to find a common denominator for a pair of fractions is to make a list of equivalent fractions.

$\frac{3}{4} = \frac{6}{8} = \frac{9}{12} = \frac{12}{16} = \frac{15}{20} = \frac{18}{24} = \ldots$ $\qquad$ $\frac{5}{6} = \frac{10}{12} = \frac{15}{18} = \frac{20}{24} = \ldots$

Another way to find a common denominator for a pair of fractions is …

____

Give the values of the variables that make each equation true.

4. $\frac{t * 4}{t * 7} = \frac{12}{21}$ $\quad$ $t =$ ________

5. $\frac{m * 4}{m * 6} = \frac{n}{30}$ $\quad$ $m =$ ____ $n =$ ____

6. $\frac{8 * x}{5 * x} = \frac{y}{45}$ $\quad$ $x =$ ____ $y =$ ____

Name a common denominator for each pair of fractions.

7. $\frac{3}{4}$ and $\frac{5}{16}$ = ________

8. $\frac{5}{8}$ and $\frac{9}{10}$ = ________

9. $\frac{4}{5}$ and $\frac{5}{6}$ = ________

Using Common Denominators

Common denominators are useful not only for adding and subtracting fractions, but also for comparing fractions.

A quick way to find a common denominator for a pair of fractions is to find the product of the denominators.

Example Compare $\frac{2}{3}$ and $\frac{5}{8}$. Use $3 * 8$ as a common denominator.

$$\frac{2}{3} = \frac{(8 * 2)}{(8 * 3)} = \frac{16}{24} \qquad \frac{5}{8} = \frac{(3 * 5)}{(3 * 8)} = \frac{15}{24}$$

$\frac{16}{24} > \frac{15}{24}$, so $\frac{2}{3} > \frac{5}{8}$.

1. Rewrite each pair of fractions below as equivalent fractions with a common denominator. Then write < (less than) or > (greater than) to compare the fractions.

	Original Fraction	Equivalent Fraction	> or <
a.	$\frac{4}{7}$		$\frac{4}{7}$ $\frac{3}{5}$
	$\frac{3}{5}$		
b.	$\frac{9}{4}$		$\frac{9}{4}$ $\frac{7}{3}$
	$\frac{7}{3}$		

Find a common denominator. Then add or subtract.

2. $\frac{1}{2} - \frac{1}{3} =$ __________

3. $\frac{7}{8} + \frac{2}{5} =$ __________

4. $\frac{3}{4} - \frac{1}{2} =$ __________

5. $\frac{4}{5} + \frac{2}{3} =$ __________

6. $\frac{9}{10} - \frac{5}{6}$

7. $\frac{1}{10} + \frac{3}{4}$

Date Time

Stem-and-Leaf Plot

1. Construct a stem-and-leaf plot with the following data landmarks. There should be at least 12 data entries in your plot.

   Median: 38 Minimum: 9 Maximum: 85 Mode: 40

2. Explain how you chose the numbers for your data set. ______________________

   ____

   ____

   ____

3. Describe a data set that your stem-and-leaf plot could represent.

   ____

   ____

   ____

Date Time

Math Boxes 6.10

1. Solve.

a. $\begin{array}{r} 28 \\ +\ 73 \\ \hline \end{array}$

b. $\begin{array}{r} 97 \\ +\ 204 \\ \hline \end{array}$

c. $\begin{array}{r} 171 \\ -\ 85 \\ \hline \end{array}$

d. $\begin{array}{r} 608 \\ -\ 321 \\ \hline \end{array}$

e. $\begin{array}{r} 1{,}752 \\ -\ 999 \\ \hline \end{array}$

2. Write a title and label the axes for the bar graph. Explain why you chose that title.

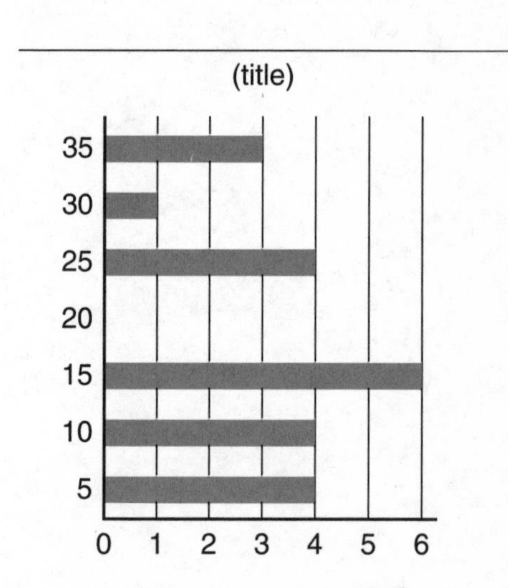

3. Use your compass and the map scale to estimate the distance from Lisa's house to Derek's house shown on the map.

The distance is about

______________.

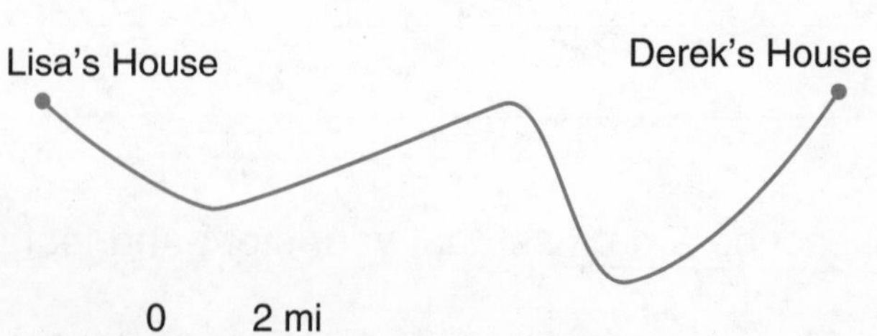

Time to Reflect

1. If you wanted to find out what the top 3 favorite TV shows of fifth graders are, about how many students would you ask? ________ students

   Explain your answer.

   ____

   ____

   ____

   ____

   ____

2. Explain one advantage of organizing data into stem-and-leaf plots.

   Example Heights in inches of Mr. Bernard's fifth grade boys

   57, 62, 64, 60, 59, 60, 57, 61, 63, 67, 59, 60

Stems (10s)	**Leaves** (1s)
5	7 7 9 9
6	0 0 0 1 2 3 4 7

   ____

   ____

   ____

   ____

   ____

   ____

Date Time

Math Boxes 6.11

1. Write a 10-digit numeral that has
8 in the ten-thousands place,
4 in the hundred-millions place,
0 in the ten-millions place,
5 in the ones place,
7 in the billions place, and
2 in all other places.

____ , ____ ____ ____ , ____ ____ ____ , ____ ____ ____

Write the numeral in words.

____

____

2. Solve.

a. $6 * 10,000 =$ ____________

b. $1,400 * 10,000 =$ ____________

c. ____________ $= 90 * 100,000$

d. ____________ $= 753 * 100,000$

e. ____________ $= 1,602 * 1,000$

3. Subtract. (*Hint:* Use a number line to help you.)

a. $5 - 15 =$ ________

b. $28 - 35 =$ ________

c. ________ $= 42 - 50$

d. ________ $= 17 - 27$

e. ________ $= 62 - 74$

4. Solve.

a. $\begin{array}{r} 229 \\ +\ 280 \\ \hline \end{array}$

b. $\begin{array}{r} 562 \\ +\ 468 \\ \hline \end{array}$

c. $\begin{array}{r} 308 \\ -\ 294 \\ \hline \end{array}$

d. $\begin{array}{r} 1,650 \\ -\ 846 \\ \hline \end{array}$

e. $\begin{array}{r} 549 \\ -\ 477 \\ \hline \end{array}$

Reference

Equivalent Fractions, Decimals, and Percents

$\frac{1}{2}$	$\frac{2}{4}$	$\frac{3}{6}$	$\frac{4}{8}$	$\frac{5}{10}$	$\frac{6}{12}$	$\frac{7}{14}$	$\frac{8}{16}$	$\frac{9}{18}$	$\frac{10}{20}$	$\frac{11}{22}$	$\frac{12}{24}$	$\frac{13}{26}$	$\frac{14}{28}$	$\frac{15}{30}$	0.5	50%
$\frac{1}{3}$	$\frac{2}{6}$	$\frac{3}{9}$	$\frac{4}{12}$	$\frac{5}{15}$	$\frac{6}{18}$	$\frac{7}{21}$	$\frac{8}{24}$	$\frac{9}{27}$	$\frac{10}{30}$	$\frac{11}{33}$	$\frac{12}{36}$	$\frac{13}{39}$	$\frac{14}{42}$	$\frac{15}{45}$	$0.\overline{3}$	$33\frac{1}{3}\%$
$\frac{2}{3}$	$\frac{4}{6}$	$\frac{6}{9}$	$\frac{8}{12}$	$\frac{10}{15}$	$\frac{12}{18}$	$\frac{14}{21}$	$\frac{16}{24}$	$\frac{18}{27}$	$\frac{20}{30}$	$\frac{22}{33}$	$\frac{24}{36}$	$\frac{26}{39}$	$\frac{28}{42}$	$\frac{30}{45}$	$0.\overline{6}$	$66\frac{2}{3}\%$
$\frac{1}{4}$	$\frac{2}{8}$	$\frac{3}{12}$	$\frac{4}{16}$	$\frac{5}{20}$	$\frac{6}{24}$	$\frac{7}{28}$	$\frac{8}{32}$	$\frac{9}{36}$	$\frac{10}{40}$	$\frac{11}{44}$	$\frac{12}{48}$	$\frac{13}{52}$	$\frac{14}{56}$	$\frac{15}{60}$	0.25	25%
$\frac{3}{4}$	$\frac{6}{8}$	$\frac{9}{12}$	$\frac{12}{16}$	$\frac{15}{20}$	$\frac{18}{24}$	$\frac{21}{28}$	$\frac{24}{32}$	$\frac{27}{36}$	$\frac{30}{40}$	$\frac{33}{44}$	$\frac{36}{48}$	$\frac{39}{52}$	$\frac{42}{56}$	$\frac{45}{60}$	0.75	75%
$\frac{1}{5}$	$\frac{2}{10}$	$\frac{3}{15}$	$\frac{4}{20}$	$\frac{5}{25}$	$\frac{6}{30}$	$\frac{7}{35}$	$\frac{8}{40}$	$\frac{9}{45}$	$\frac{10}{50}$	$\frac{11}{55}$	$\frac{12}{60}$	$\frac{13}{65}$	$\frac{14}{70}$	$\frac{15}{75}$	0.2	20%
$\frac{2}{5}$	$\frac{4}{10}$	$\frac{6}{15}$	$\frac{8}{20}$	$\frac{10}{25}$	$\frac{12}{30}$	$\frac{14}{35}$	$\frac{16}{40}$	$\frac{18}{45}$	$\frac{20}{50}$	$\frac{22}{55}$	$\frac{24}{60}$	$\frac{26}{65}$	$\frac{28}{70}$	$\frac{30}{75}$	0.4	40%
$\frac{3}{5}$	$\frac{6}{10}$	$\frac{9}{15}$	$\frac{12}{20}$	$\frac{15}{25}$	$\frac{18}{30}$	$\frac{21}{35}$	$\frac{24}{40}$	$\frac{27}{45}$	$\frac{30}{50}$	$\frac{33}{55}$	$\frac{36}{60}$	$\frac{39}{65}$	$\frac{42}{70}$	$\frac{45}{75}$	0.6	60%
$\frac{4}{5}$	$\frac{8}{10}$	$\frac{12}{15}$	$\frac{16}{20}$	$\frac{20}{25}$	$\frac{24}{30}$	$\frac{28}{35}$	$\frac{32}{40}$	$\frac{36}{45}$	$\frac{40}{50}$	$\frac{44}{55}$	$\frac{48}{60}$	$\frac{52}{65}$	$\frac{56}{70}$	$\frac{60}{75}$	0.8	80%
$\frac{1}{6}$	$\frac{2}{12}$	$\frac{3}{18}$	$\frac{4}{24}$	$\frac{5}{30}$	$\frac{6}{36}$	$\frac{7}{42}$	$\frac{8}{48}$	$\frac{9}{54}$	$\frac{10}{60}$	$\frac{11}{66}$	$\frac{12}{72}$	$\frac{13}{78}$	$\frac{14}{84}$	$\frac{15}{90}$	$0.1\overline{6}$	$16\frac{2}{3}\%$
$\frac{5}{6}$	$\frac{10}{12}$	$\frac{15}{18}$	$\frac{20}{24}$	$\frac{25}{30}$	$\frac{30}{36}$	$\frac{35}{42}$	$\frac{40}{48}$	$\frac{45}{54}$	$\frac{50}{60}$	$\frac{55}{66}$	$\frac{60}{72}$	$\frac{65}{78}$	$\frac{70}{84}$	$\frac{75}{90}$	$0.8\overline{3}$	$83\frac{1}{3}\%$
$\frac{1}{7}$	$\frac{2}{14}$	$\frac{3}{21}$	$\frac{4}{28}$	$\frac{5}{35}$	$\frac{6}{42}$	$\frac{7}{49}$	$\frac{8}{56}$	$\frac{9}{63}$	$\frac{10}{70}$	$\frac{11}{77}$	$\frac{12}{84}$	$\frac{13}{91}$	$\frac{14}{98}$	$\frac{15}{105}$	0.143	14.3%
$\frac{2}{7}$	$\frac{4}{14}$	$\frac{6}{21}$	$\frac{8}{28}$	$\frac{10}{35}$	$\frac{12}{42}$	$\frac{14}{49}$	$\frac{16}{56}$	$\frac{18}{63}$	$\frac{20}{70}$	$\frac{22}{77}$	$\frac{24}{84}$	$\frac{26}{91}$	$\frac{28}{98}$	$\frac{30}{105}$	0.286	28.6%
$\frac{3}{7}$	$\frac{6}{14}$	$\frac{9}{21}$	$\frac{12}{28}$	$\frac{15}{35}$	$\frac{18}{42}$	$\frac{21}{49}$	$\frac{24}{56}$	$\frac{27}{63}$	$\frac{30}{70}$	$\frac{33}{77}$	$\frac{36}{84}$	$\frac{39}{91}$	$\frac{42}{98}$	$\frac{45}{105}$	0.429	42.9%
$\frac{4}{7}$	$\frac{8}{14}$	$\frac{12}{21}$	$\frac{16}{28}$	$\frac{20}{35}$	$\frac{24}{42}$	$\frac{28}{49}$	$\frac{32}{56}$	$\frac{36}{63}$	$\frac{40}{70}$	$\frac{44}{77}$	$\frac{48}{84}$	$\frac{52}{91}$	$\frac{56}{98}$	$\frac{60}{105}$	0.571	57.1%
$\frac{5}{7}$	$\frac{10}{14}$	$\frac{15}{21}$	$\frac{20}{28}$	$\frac{25}{35}$	$\frac{30}{42}$	$\frac{35}{49}$	$\frac{40}{56}$	$\frac{45}{63}$	$\frac{50}{70}$	$\frac{55}{77}$	$\frac{60}{84}$	$\frac{65}{91}$	$\frac{70}{98}$	$\frac{75}{105}$	0.714	71.4%
$\frac{6}{7}$	$\frac{12}{14}$	$\frac{18}{21}$	$\frac{24}{28}$	$\frac{30}{35}$	$\frac{36}{42}$	$\frac{42}{49}$	$\frac{48}{56}$	$\frac{54}{63}$	$\frac{60}{70}$	$\frac{66}{77}$	$\frac{72}{84}$	$\frac{78}{91}$	$\frac{84}{98}$	$\frac{90}{105}$	0.857	85.7%
$\frac{1}{8}$	$\frac{2}{16}$	$\frac{3}{24}$	$\frac{4}{32}$	$\frac{5}{40}$	$\frac{6}{48}$	$\frac{7}{56}$	$\frac{8}{64}$	$\frac{9}{72}$	$\frac{10}{80}$	$\frac{11}{88}$	$\frac{12}{96}$	$\frac{13}{104}$	$\frac{14}{112}$	$\frac{15}{120}$	0.125	$12\frac{1}{2}\%$
$\frac{3}{8}$	$\frac{6}{16}$	$\frac{9}{24}$	$\frac{12}{32}$	$\frac{15}{40}$	$\frac{18}{48}$	$\frac{21}{56}$	$\frac{24}{64}$	$\frac{27}{72}$	$\frac{30}{80}$	$\frac{33}{88}$	$\frac{36}{96}$	$\frac{39}{104}$	$\frac{42}{112}$	$\frac{45}{120}$	0.375	$37\frac{1}{2}\%$
$\frac{5}{8}$	$\frac{10}{16}$	$\frac{15}{24}$	$\frac{20}{32}$	$\frac{25}{40}$	$\frac{30}{48}$	$\frac{35}{56}$	$\frac{40}{64}$	$\frac{45}{72}$	$\frac{50}{80}$	$\frac{55}{88}$	$\frac{60}{96}$	$\frac{65}{104}$	$\frac{70}{112}$	$\frac{75}{120}$	0.625	$62\frac{1}{2}\%$
$\frac{7}{8}$	$\frac{14}{16}$	$\frac{21}{24}$	$\frac{28}{32}$	$\frac{35}{40}$	$\frac{42}{48}$	$\frac{49}{56}$	$\frac{56}{64}$	$\frac{63}{72}$	$\frac{70}{80}$	$\frac{77}{88}$	$\frac{84}{96}$	$\frac{91}{104}$	$\frac{98}{112}$	$\frac{105}{120}$	0.875	$87\frac{1}{2}\%$
$\frac{1}{9}$	$\frac{2}{18}$	$\frac{3}{27}$	$\frac{4}{36}$	$\frac{5}{45}$	$\frac{6}{54}$	$\frac{7}{63}$	$\frac{8}{72}$	$\frac{9}{81}$	$\frac{10}{90}$	$\frac{11}{99}$	$\frac{12}{108}$	$\frac{13}{117}$	$\frac{14}{126}$	$\frac{15}{135}$	$0.\overline{1}$	$11\frac{1}{9}\%$
$\frac{2}{9}$	$\frac{4}{18}$	$\frac{6}{27}$	$\frac{8}{36}$	$\frac{10}{45}$	$\frac{12}{54}$	$\frac{14}{63}$	$\frac{16}{72}$	$\frac{18}{81}$	$\frac{20}{90}$	$\frac{22}{99}$	$\frac{24}{108}$	$\frac{26}{117}$	$\frac{28}{126}$	$\frac{30}{135}$	$0.\overline{2}$	$22\frac{2}{9}\%$
$\frac{4}{9}$	$\frac{8}{18}$	$\frac{12}{27}$	$\frac{16}{36}$	$\frac{20}{45}$	$\frac{24}{54}$	$\frac{28}{63}$	$\frac{32}{72}$	$\frac{36}{81}$	$\frac{40}{90}$	$\frac{44}{99}$	$\frac{48}{108}$	$\frac{52}{117}$	$\frac{56}{126}$	$\frac{60}{135}$	$0.\overline{4}$	$44\frac{4}{9}\%$
$\frac{5}{9}$	$\frac{10}{18}$	$\frac{15}{27}$	$\frac{20}{36}$	$\frac{25}{45}$	$\frac{30}{54}$	$\frac{35}{63}$	$\frac{40}{72}$	$\frac{45}{81}$	$\frac{50}{90}$	$\frac{55}{99}$	$\frac{60}{108}$	$\frac{65}{117}$	$\frac{70}{126}$	$\frac{75}{135}$	$0.\overline{5}$	$55\frac{5}{9}\%$
$\frac{7}{9}$	$\frac{14}{18}$	$\frac{21}{27}$	$\frac{28}{36}$	$\frac{35}{45}$	$\frac{42}{54}$	$\frac{49}{63}$	$\frac{56}{72}$	$\frac{63}{81}$	$\frac{70}{90}$	$\frac{77}{99}$	$\frac{84}{108}$	$\frac{91}{117}$	$\frac{98}{126}$	$\frac{105}{135}$	$0.\overline{7}$	$77\frac{7}{9}\%$
$\frac{8}{9}$	$\frac{16}{18}$	$\frac{24}{27}$	$\frac{32}{36}$	$\frac{40}{45}$	$\frac{48}{54}$	$\frac{56}{63}$	$\frac{64}{72}$	$\frac{72}{81}$	$\frac{80}{90}$	$\frac{88}{99}$	$\frac{96}{108}$	$\frac{104}{117}$	$\frac{112}{126}$	$\frac{120}{135}$	$0.\overline{8}$	$88\frac{8}{9}\%$

Note: The decimals for sevenths have been rounded to the nearest thousandth.

Reference

Metric System

Units of Length

1 kilometer (km)	=	1000 meters (m)
1 meter	=	10 decimeters (dm)
	=	100 centimeters (cm)
	=	1000 millimeters (mm)
1 decimeter	=	10 centimeters
1 centimeter	=	10 millimeters

Units of Area

1 square meter (m^2)	=	100 square decimeters (dm^2)
	=	10,000 square centimeters (cm^2)
1 square decimeter	=	100 square centimeters
1 are (a)	=	100 square meters
1 hectare (ha)	=	100 ares
1 square kilometer (km^2)	=	100 hectares

Units of Volume

1 cubic meter (m^3)	=	1000 cubic decimeters (dm^3)
	=	1,000,000 cubic centimeters (cm^3)
1 cubic decimeter	=	1000 cubic centimeters

Units of Capacity

1 kiloliter (kL)	=	1000 liters (L)
1 liter	=	1000 milliliters (mL)

Units of Mass

1 metric ton (t)	=	1000 kilograms (kg)
1 kilogram	=	1000 grams (g)
1 gram	=	1000 milligrams (mg)

Units of Time

1 century	=	100 years
1 decade	=	10 years
1 year (yr)	=	12 months
	=	52 weeks (plus one or two days)
	=	365 days (366 days in a leap year)
1 month (mo)	=	28, 29, 30, or 31 days
1 week (wk)	=	7 days
1 day (d)	=	24 hours
1 hour (hr)	=	60 minutes
1 minute (min)	=	60 seconds (sec)

U.S. Customary System

Units of Length

1 mile (mi)	=	1760 yards (yd)
	=	5280 feet (ft)
1 yard	=	3 feet
	=	36 inches (in.)
1 foot	=	12 inches

Units of Area

1 square yard (yd^2)	=	9 square feet (ft^2)
	=	1296 square inches ($in.^2$)
1 square foot	=	144 square inches
1 acre	=	43,560 square feet
1 square mile (mi^2)	=	640 acres

Units of Volume

1 cubic yard (yd^3)	=	27 cubic feet (ft^3)
1 cubic foot	=	1728 cubic inches ($in.^3$)

Units of Capacity

1 gallon (gal)	=	4 quarts (qt)
1 quart	=	2 pints (pt)
1 pint	=	2 cups (c)
1 cup	=	8 fluid ounces (fl oz)
1 fluid ounce	=	2 tablespoons (tbs)
1 tablespoon	=	3 teaspoons (tsp)

Units of Weight

1 ton (T)	=	2000 pounds (lb)
1 pound	=	16 ounces (oz)

System Equivalents

1 inch is about 2.5 cm (2.54)

1 kilometer is about 0.6 mile (0.621)

1 mile is about 1.6 kilometers (1.609)

1 meter is about 39 inches (39.37)

1 liter is about 1.1 quarts (1.057)

1 ounce is about 28 grams (28.350)

1 kilogram is about 2.2 pounds (2.205)

1 hectare is about 2.5 acres (2.47)

Rules for Order of Operations

1. Do operations within parentheses or other grouping symbols before doing anything else.
2. Calculate all powers.
3. Do multiplications or divisions in order, from left to right.
4. Then do additions or subtractions in order, from left to right.

Reference

Place-Value Chart

trillions	100B	10B	billions	100M	10M	millions	hundred-thousands	ten-thousands	thousands	hundreds	tens	ones	.	tenths	hundredths	thousandths
1000 billions			1000 millions			1,000,000s	100,000s	10,000s	1000s	100s	10s	1s	.	0.1s	0.01s	0.001s
10^{12}	10^{11}	10^{10}	10^{9}	10^{8}	10^{7}	10^{6}	10^{5}	10^{4}	10^{3}	10^{2}	10^{1}	10^{0}	.	10^{-1}	10^{-2}	10^{-3}

Probability Meter

CERTAIN

Percent	Decimal	Fraction	
100%	1.00	1	
	0.99	$\frac{99}{100}$	EXTREMELY LIKELY
95%	0.95	$\frac{19}{20}$	
90%	0.90	$\frac{9}{10}$	VERY LIKELY
	0.875	$\frac{7}{8}$	
85%	0.85		
	$0.8\overline{3}$	$\frac{5}{6}$	
80%	0.80	$\frac{4}{5}, \frac{8}{10}$	
75%	0.75	$\frac{3}{4}, \frac{6}{8}$	LIKELY
70%	0.70	$\frac{7}{10}$	
	$0.6\overline{6}$	$\frac{2}{3}$	
65%	0.65		
	0.625	$\frac{5}{8}$	
60%	0.60	$\frac{3}{5}, \frac{6}{10}$	
55%	0.55		
50%	0.50	$\frac{1}{2}, \frac{2}{4}, \frac{3}{6}, \frac{4}{8}, \frac{5}{10}, \frac{10}{20}, \frac{50}{100}$	50-50 CHANCE
45%	0.45		
40%	0.40	$\frac{2}{5}, \frac{4}{10}$	
	0.375	$\frac{3}{8}$	
35%	0.35		UNLIKELY
	$0.3\overline{3}$	$\frac{1}{3}$	
30%	0.30	$\frac{3}{10}$	
25%	0.25	$\frac{1}{4}, \frac{2}{8}$	
20%	0.20	$\frac{1}{5}$	VERY UNLIKELY
	$0.1\overline{6}$	$\frac{1}{6}$	
15%	0.15		
	0.125	$\frac{1}{8}$	
10%	0.10	$\frac{1}{10}$	
5%	0.05	$\frac{1}{20}$	EXTREMELY UNLIKELY
	0.01	$\frac{1}{100}$	
0%	0.00	0	

IMPOSSIBLE

Symbols

+	plus or positive
−	minus or negative
∗, ×	multiplied by
÷, /	divided by
=	is equal to
≠	is not equal to
<	is less than
>	is greater than
≤	is less than or equal to
≥	is greater than or equal to
x^n	nth power of x
$\sqrt{x}$	square root of x
%	percent
$a{:}b$, a/b, $\frac{a}{b}$	ratio of a to b or a divided by b or the fraction $\frac{a}{b}$
°	degree
(a,b)	ordered pair
$\overleftrightarrow{AS}$	line AS
$\overline{AS}$	line segment AS
$\overrightarrow{AS}$	ray AS
∟	right angle
⊥	is perpendicular to
∥	is parallel to
$\triangle ABC$	triangle ABC
$\angle ABC$	angle ABC
$\angle B$	angle B

Latitude and Longitude

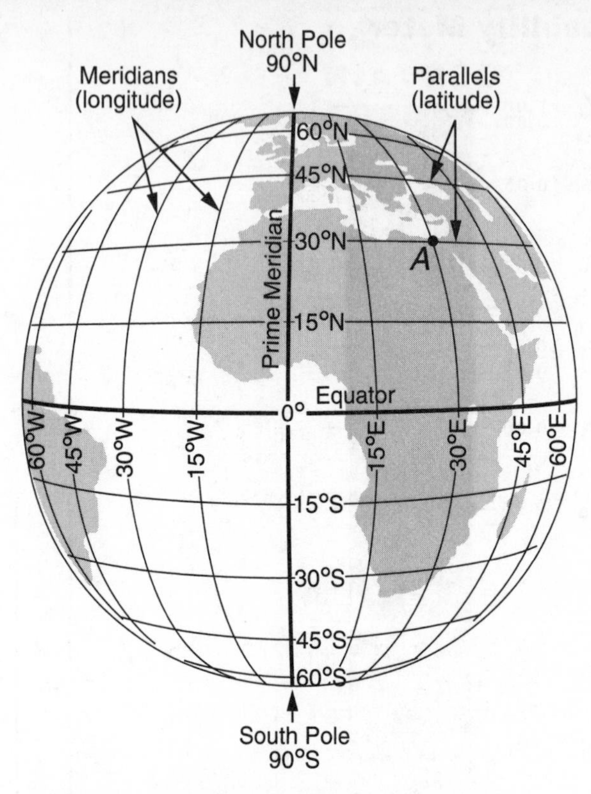

Point *A* is located at 30°N latitude and 30°E longitude.

Fraction-Stick and Decimal Number-Line Chart

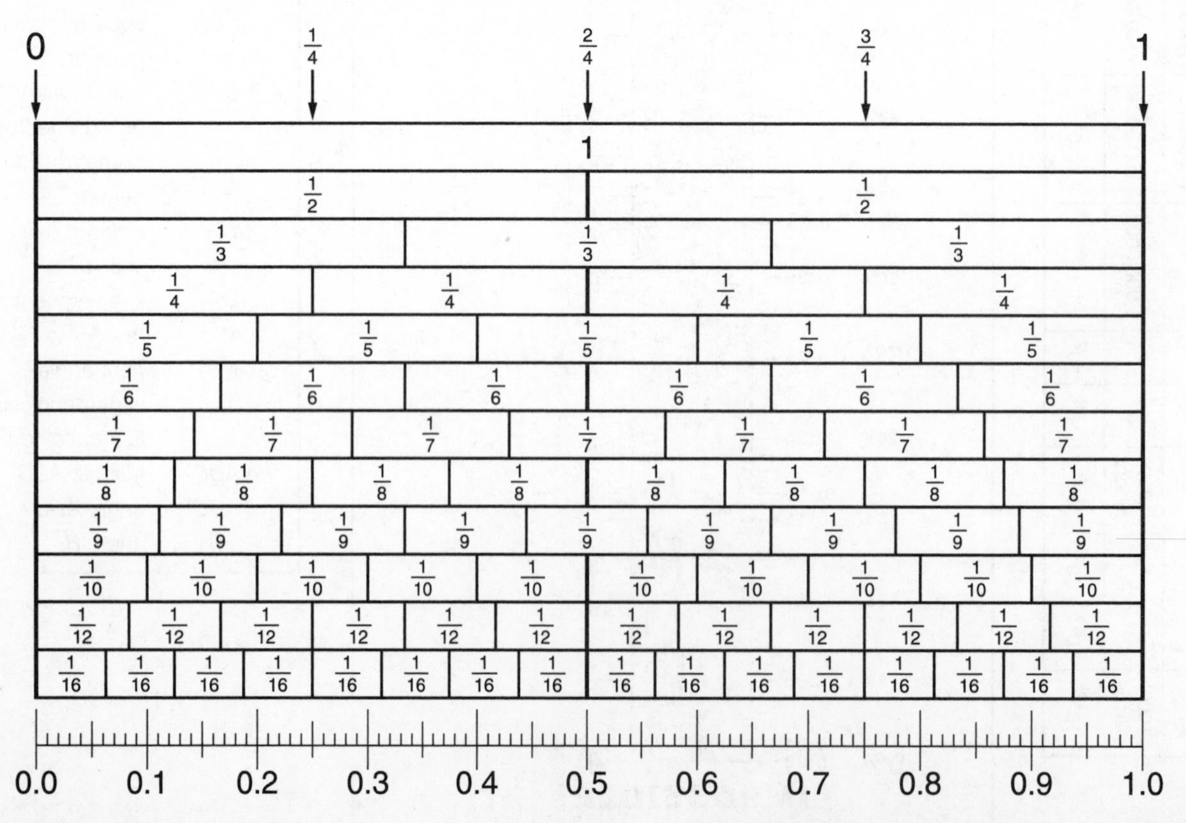

Date Time

*,/ Fact Triangles

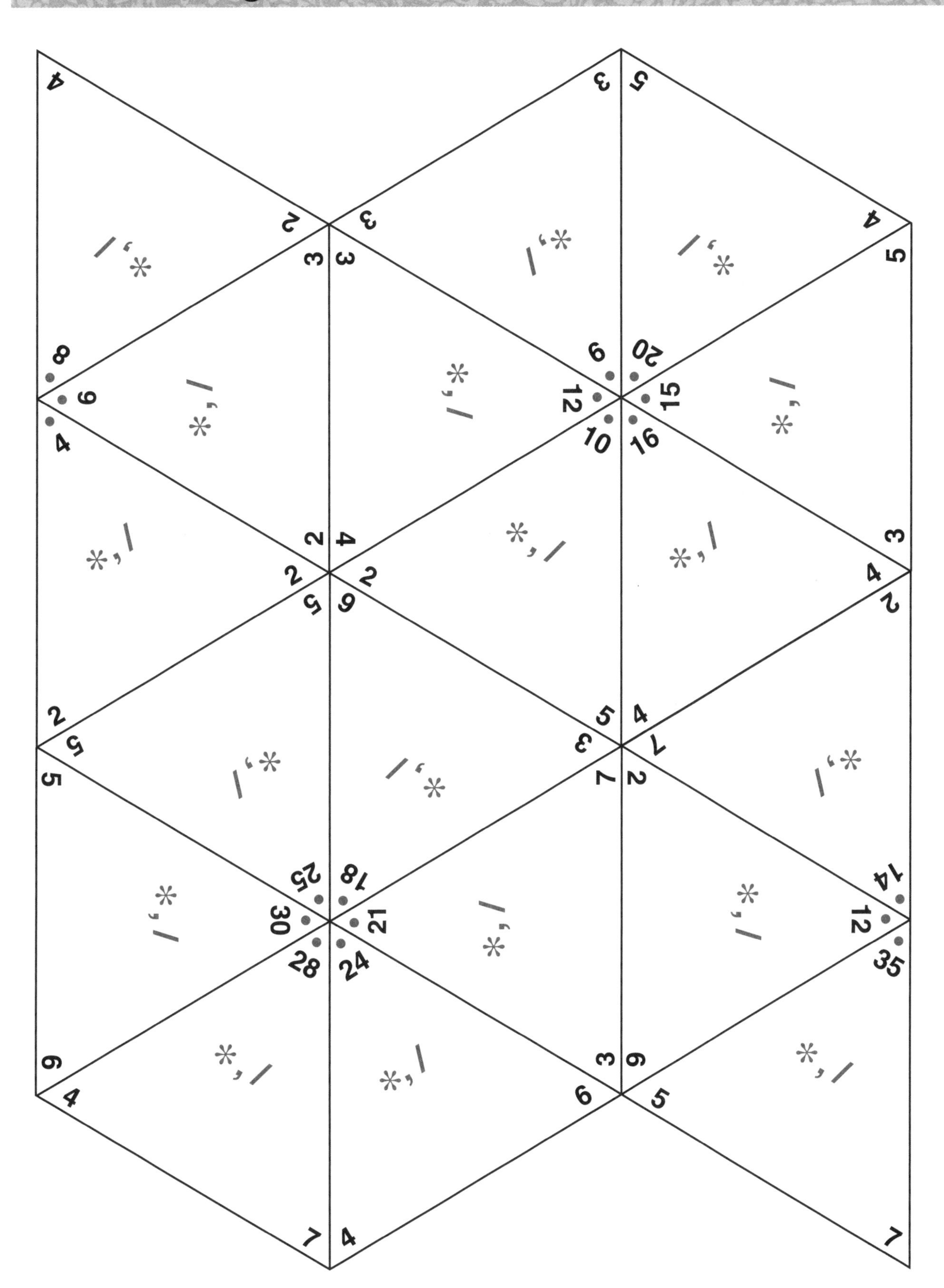

Date Time

*,/ Fact Triangles

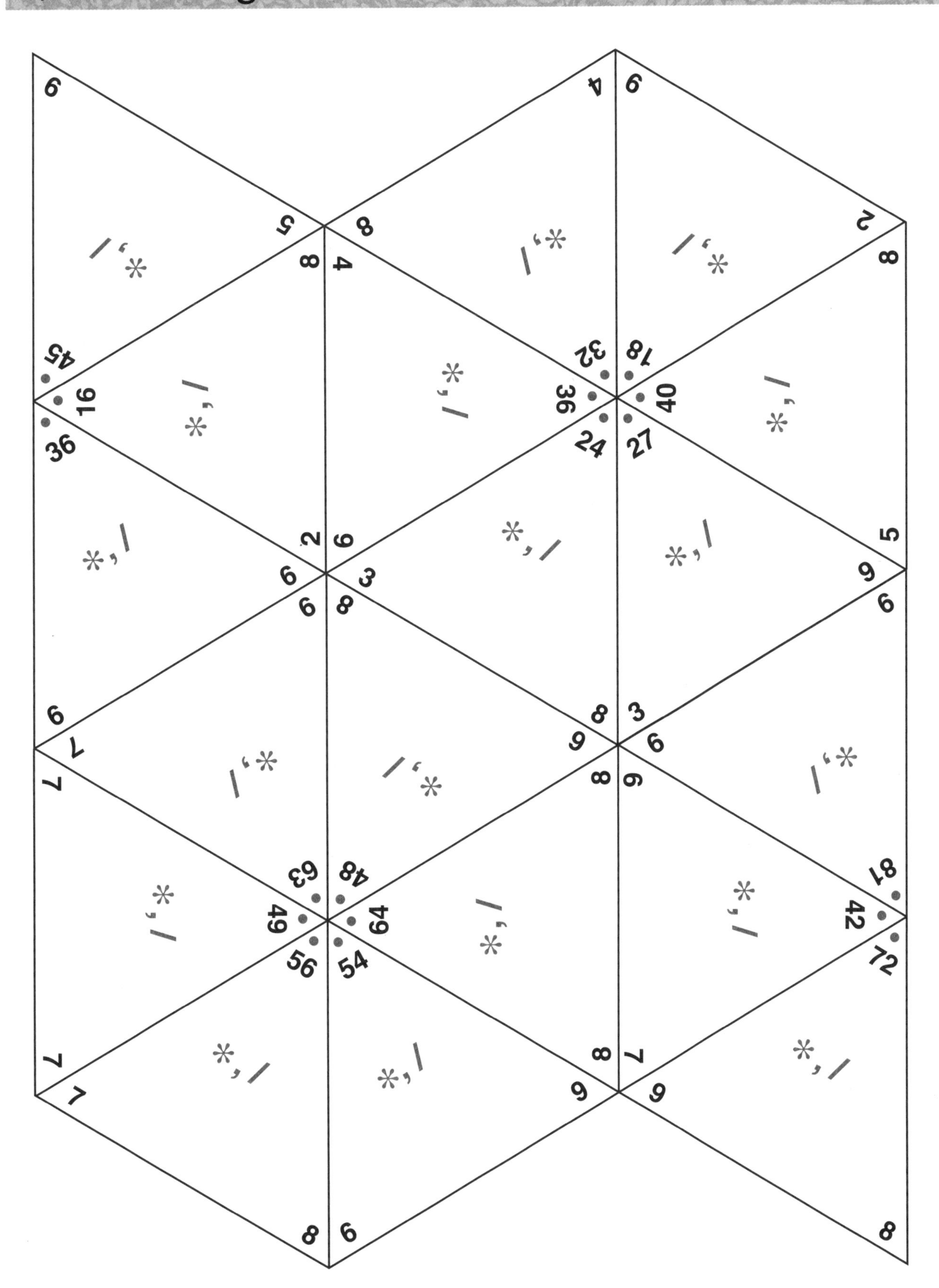

Date Time

Grab-It-Gauge

0.22 second

0.21

0.20

0.19

0.18

0.17

0.16

0.15

0.14

0.13

0.12

0.11

0.10

0.09

0.08

0.07

0.00 **starting position for contestant**

Date Time

Polygon Capture Pieces

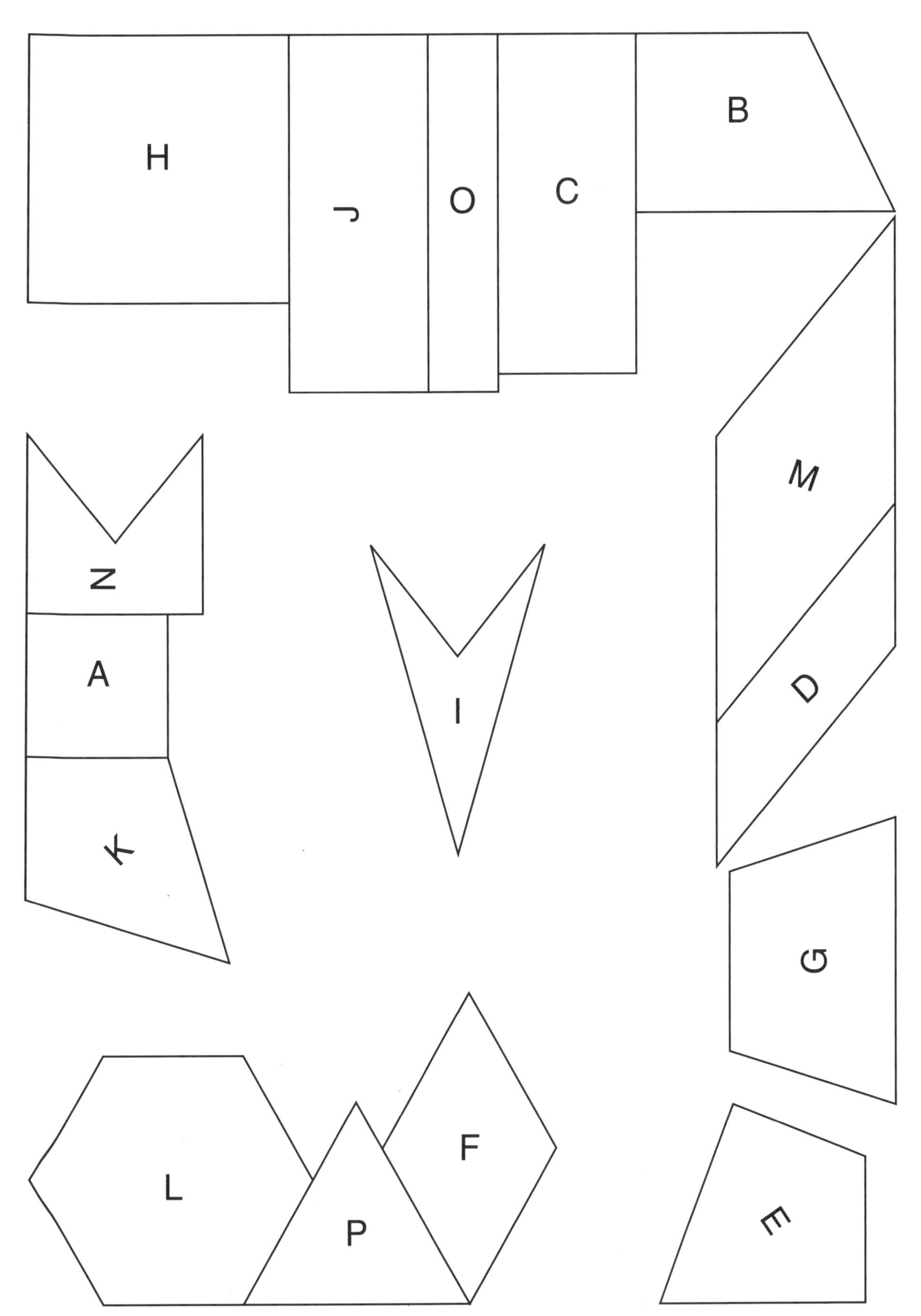